Lecture Notes in Artificial Intelligence   12588

Subseries of Lecture Notes in Computer Science

More information about this subseries at http://www.springer.com/series/1244

Vincent Lemaire · Simon Malinowski ·
Anthony Bagnall · Thomas Guyet ·
Romain Tavenard · Georgiana Ifrim (Eds.)

Advanced Analytics and Learning on Temporal Data

5th ECML PKDD Workshop, AALTD 2020
Ghent, Belgium, September 18, 2020
Revised Selected Papers

Springer

Editors
Vincent Lemaire ⓘ
Orange Labs
Lannion, France

Anthony Bagnall ⓘ
University of East Anglia
Norwich, UK

Romain Tavenard ⓘ
CNRS, LETG/IRISA
University of Rennes 2
Rennes, France

Simon Malinowski ⓘ
Inria
University of Rennes
Rennes, France

Thomas Guyet ⓘ
Agrocampus Ouest/IRISA
Rennes, France

Georgiana Ifrim ⓘ
University College Dublin
Dublin, Ireland

ISSN 0302-9743   ISSN 1611-3349   (electronic)
Lecture Notes in Artificial Intelligence
ISBN 978-3-030-65741-3   ISBN 978-3-030-65742-0   (eBook)
https://doi.org/10.1007/978-3-030-65742-0

LNCS Sublibrary: SL7 – Artificial Intelligence

This Springer imprint is published by the registered company Springer Nature Switzerland AG
The registered company address is: Gewerbestrasse 11, 6330 Cham, Switzerland

Preface

Workshop Description

The European Conference on Machine Learning and Principles and Practice of Knowledge Discovery in Databases (ECML-PKDD) is the premier European machine learning and data mining conference and builds upon over 18 years of successful events and conferences held across Europe. This year, ECML-PKDD 2020, was planned to take place in Ghent, Belgium, during September 14–18, 2020, but due to the COVID-19 pandemic it was held in the same time period as a fully virtual event. The main conference was complemented by a workshop program, where each workshop was dedicated to specialized topics, cross-cutting issues, and upcoming research trends. This standalone LNAI volume includes the selected papers of the 5th Workshop on Advanced Analytics and Learning on Temporal Data (AALTD) held at ECML-PKDD 2020.

Motivation – Temporal data are frequently encountered in a wide range of domains such as bio-informatics, medicine, finance, and engineering, among many others. They are naturally present in emerging applications such as motion analysis, energy efficient building, smart cities, dynamic social media, or sensor networks. Contrary to static data, temporal data are of complex nature, they are generally noisy, of high dimensionality, they may be non stationary (i.e. first order statistics vary with time) and irregular (i.e. involving several time granularities) and they may have several invariant domain-dependent factors such as time delay, translation, scale, or tendency effects. These temporal peculiarities limit the majority of standard statistical models and machine learning approaches, that mainly assume i.i.d data, homoscedasticity, normality of residuals, etc. To tackle such challenging temporal data, one appeals for new advanced approaches at the bridge of statistics, time series analysis, signal processing, and machine learning. Defining new approaches that transcend boundaries between several domains to extract valuable information from temporal data is undeniably a hot topic and it has been the subject of active research this last decade.

Workshop Topics – The aim of the workshop series on AALTD[1] was to bring together researchers and experts in machine learning, data mining, pattern analysis, and statistics to share their challenging issues and advance in temporal data analysis. Analysis and learning from temporal data covers a wide scope of tasks including learning metrics, learning representations, unsupervised feature extraction, clustering, and classification.

---

[1] https://project.inria.fr/aaltd20/.

For this fourth edition, the proposed workshop received papers that cover one or several of the following topics:

- Temporal Data Clustering
- Classification of Univariate and Multivariate Time Series
- Early Classification of Temporal Data
- Deep Learning and Learning Representations for Temporal Data
- Modeling Temporal Dependencies
- Advanced Forecasting and Prediction Models
- Space-Temporal Statistical Analysis
- Functional Data Analysis Methods
- Temporal Data Streams
- Interpretable Time-Series Analysis Methods
- Dimensionality Reduction, Sparsity, Algorithmic Complexity, and Big Data Challenge
- Bio-Informatics, Medical, Energy Consumption, on Temporal Data

Outcomes – AALTD 2020 was structured as a full-day workshop. We encouraged submissions of regular papers that were up to 16 pages of previously unpublished work. All submitted papers were peer reviewed (double-blind) by two or three reviewers from the Program Committee, and selected on the basis of these reviews. AALTD 2020 received 29 submissions, among which 15 papers were accepted for inclusion in the proceedings. The papers with the highest review rating were selected for oral presentation, and the others were given the opportunity to present a poster through a spotlight session and a discussion session. The workshop had an invited talk "Scalable Machine Learning on Large Sequence Collections"[2] given by Professor Themis Palpanas of the French University Institute (IUF) and University of Paris, France.

We thank all organizers, reviewers, and authors for the time and effort invested to make this workshop a success. We would also like to express our gratitude to the members of the Program Committee. We thank the Organizing Committee of ECML-PKDD 2020 and the technical staff who helped us to make the virtual AALTD a successful workshop. Sincere thanks are due to Springer for their help in publishing the proceedings. Lastly, we thank all participants and speakers at AALTD 2020 for their contributions, their collective support has made the workshop a really interesting and successful event, even under the challenging circumstances of a global pandemic.

November 2020

Vincent Lemaire
Simon Malinowski
Anthony Bagnall
Thomas Guyet
Romain Tavenard
Georgiana Ifrim

---

[2] https://project.inria.fr/aaltd20/invited-speaker/.

Organization

Program Committee Chairs

Anthony Bagnall	University of East Anglia, UK
Thomas Guyet	Institute Agro, IRISA, France
Georgiana Ifrim	University College Dublin, Ireland
Vincent Lemaire	Orange Labs, France
Simon Malinowski	Université de Rennes, Inria, CNRS, IRISA, France
Romain Tavenard	Université de Rennes 2, COSTEL, France

Program Committee

Amaia Abanda	Basque Center for Applied Mathematics (BCAM), Spain
Mustafa Baydoğan	Boğaziçi University, Turkey
Alexis Bondu	Orange Labs, France
Antoine Cornuéjols	AgroParisTech, France
Padraig Cunningham	University College Dublin, Ireland
Elias Egho	Orange Labs, France
Germain Forestier	Université de Haute-Alsace, France
Dominique Gay	Université de La Réunion, France
Severin Gsponer	Insight Centre for Data Analytics, Ireland
David Guijo-Rubio	Universidad de Córdoba, Spain
Neil Hurley	University College Dublin, Ireland
Hassan Ismail Fawaz	Université de Haute-Alsace, France
James Large	University of East Anglia, UK
Jason Lines	University of East Anglia, UK
Brian Mac Namee	University College Dublin, Ireland
Usue Mori	University of the Basque Country, Spain
Charlotte Pelletier	Université de Bretagne-Sud, IRISA, France
Patrick Schäfer	Humboldt-Universität zu Berlin, Germany
Pavel Senin	Los Alamos National Laboratory, USA
Diego Silva	Universidade Federal de São Carlos, Brazil
Chang Wei	Monash University, Australia
Julien Velcin	ERIC, Université Lyon 2, France

Contents

Oral Presentation

On the Usage and Performance of the Hierarchical Vote Collective of Transformation-Based Ensembles Version 1.0 (HIVE-COTE v1.0)

Anthony Bagnall[✉], Michael Flynn, James Large, Jason Lines, and Matthew Middlehurst

University of East Anglia, Norwich, UK
ajb@uea.ac.uk

Abstract. The Hierarchical Vote Collective of Transformation-based Ensembles (HIVE-COTE) is a heterogeneous meta ensemble for time series classification. Since it was first proposed in 2016, the algorithm has undergone some minor changes and there is now a configurable, scalable and easy to use version available in two open source repositories. We present an overview of the latest stable HIVE-COTE, version 1.0, and describe how it differs to the original. We provide a walkthrough guide of how to use the classifier, and conduct extensive experimental evaluation of its predictive performance and resource usage. We compare the performance of HIVE-COTE to three recently proposed algorithms.

Keywords: Time series · Classification · Heterogeneous ensembles · HIVE-COTE

1 Introduction

The Hierarchical Vote Collective of Transformation-based Ensembles (HIVE-COTE) is a heterogeneous meta ensemble for time series classification. The key principle behind HIVE-COTE is that time series classification (TSC) problems are best approached by careful consideration of the data representation, and that with no expert knowledge to the contrary, the most accurate algorithm design is to ensemble classifiers built on different representations.

HIVE-COTE was first described in 2016 [14,15]. At the time, HIVE-COTE was significantly more accurate on average than other known approaches [3] on the 85 datasets that were then the complete UCR archive [6]. HIVE-COTE was an improvement over the 2015 version, called just COTE on publication [4] but later renamed Flat-COTE to differentiate it from its successor. Flat-COTE is a standard ensemble of a range of classifiers built on different representations. It was itself a natural extension of the Elastic Ensemble [13] which only contains nearest neighbour classifiers using different distance measures. HIVE-COTE takes a more structured approach than Flat-COTE. The original HIVE-COTE, which we will henceforth refer to as HIVE-COTE alpha, contained the

© Springer Nature Switzerland AG 2020
V. Lemaire et al. (Eds.): AALTD 2020, LNAI 12588, pp. 3–18, 2020.
https://doi.org/10.1007/978-3-030-65742-0_1

following classification modules: Elastic Ensemble (EE) [13]; Shapelet Transform Classifier (STC) [11]; Time Series Forest (TSF) [8]; and Bag of Symbolic-Fourier-Approximation Symbols (BOSS) [19]. Each module is encapsulated and built on the train data independently of the others. For new data, each module passes an estimate of class probabilities to the control unit, which combines them to form a single prediction. It does this by weighting the probabilities of each module by an estimate of its testing accuracy formed from the training data.

Our goal with HIVE-COTE alpha was to achieve the highest level of accuracy without concern for the computational resources. This has lead to the perception that HIVE-COTE is very slow and does not scale well. Whilst this is true if used in its basic form, it is in fact very simple to restructure HIVE-COTE so it achieves the same level of accuracy in orders of magnitude less time. We have made many small changes to HIVE-COTE with the goal of making it scalable and more useful. We describe these improvements and encapsulate them as HIVE-COTE version 1.0. The changes in HIVE-COTE are both algorithmic and engineering in nature.

The two slowest components of HIVE-COTE alpha are STC (in its old format) and EE. STC used to conduct a full enumeration of all possible shapelets. We have found that this enormous computational effort is not only unnecessary, but often results in over-fitting. EE requires cross validation of numerous nearest neighbour classifiers and is very slow on training and testing. EE resulted from a comparative study of nearest neighbour distance measures. Our hypothesis was there was no significant difference between the numerous distance measures and dynamic time warping when used in nearest neighbour classifiers, which is true. We only ensembled as an afterthought. We were surprised to see significant improvement. Its design was necessarily ad hoc to avoid over-fitting. We have found that dropping EE all together does not make HIVE-COTE much worse. The main changes are:

1. STC no longer fully enumerates the shapelet space.
2. EE is dropped altogether from HIVE-COTE.
3. The STC, BOSS and RISE components include revisions to improve efficiency.
4. All components and HIVE-COTE 1.0 itself are now contractable (you can set a run time limit), checkpointable (you can save a version to continue building later) and tuneable (select parameters based on train set performance).
5. HIVE-COTE 1.0 can be threaded, built from existing results and easily configured.

The aim of this report is to describe the changes, showcase the usage of HIVE-COTE and to present some new benchmark results that should be used in all future experiments, and to demonstrate the scalability of HIVE-COTE.

2   HIVE-COTE 1.0 Design

Figure 1 provides an overview of the HIVE-COTE structure. The top level ensemble structure and the implementation of each component are described below.

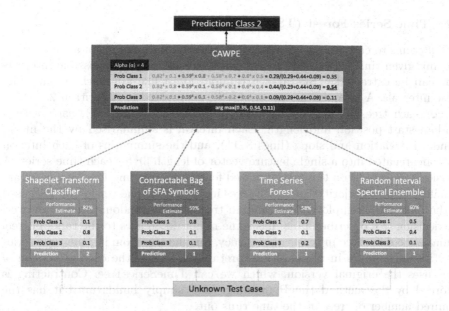

Fig. 1. An overview of the ensemble structure of HIVE-COTE 1.0. Each module produces an estimate of the probability of membership of each class. The control unit (CAWPE) combines these probabilities, weighted by an estimate of the quality of the module found on the train data.

2.1 Ensemble Structure

HIVE-COTE adopts the Cross-validation Accuracy Weighted Probabilistic Ensemble (CAWPE) ensemble structure [12], summarised in Algorithm 1. CAWPE uses an estimate of the accuracy of each classifier to weight the probability estimates of each component. It constructs a tilted distribution through exponentiation (using α) to extenuate differences in classifiers. The weight for each component is found either through ten fold cross validation, or, if the classifier has the capability to estimate its own performance, internally.

Algorithm 1. HIVE-COTE classify(A test case $\mathbf{x}$)

Return: prediction for case $\mathbf{x}$
Parameters: A set of classifiers $< M_1, \ldots, M_k >$, an exponent α, a set of weights w_i, and the number of classes c
1: $\{\hat{p}_1, \ldots, \hat{p}_c\} = \{0, \ldots, 0\}$
2: **for** $i \leftarrow 1$ to k **do**
3:    **for** $j \leftarrow 1$ to c **do**
4:       $\hat{q}_j \leftarrow \hat{p}((y = j | M_i, \mathbf{x})$
5:       $\hat{p}_j \leftarrow \hat{p}_j + w_i^\alpha \cdot \hat{q}_j$
6: **return** $argmax_{j=1\ldots c}\hat{p}_j$

2.2   Time Series Forest (TSF)

TSF [8] aims to capture basic summary features from intervals of a time series. For any given time series of length m there are $m(m-1)/2$ possible intervals that can be extracted. TSF takes a random forest-like approach to sampling these intervals. A formal description of TSF is provided in Algorithm 2.

For each tree, r intervals are randomly selected (lines 5–7), each with a random start position and length. Each interval is summarised by the mean, standard deviation and slope (lines 8–11), and the summaries of each interval are concatenated into a single feature vector of length $3r$ for each time series. A decision tree is built on this concatenated feature vector (line 12). New cases are classified using a majority vote of all trees in the forest. The version of TSF used in the bake off [3] employed the random tree used by random forest. However, the decision tree described in [8] has some minor differences to the random tree. It makes no difference in terms of accuracy, but the tree from [8], the time series tree, has advantages in terms of interpretability. Hence, the current version of TSF uses the original version, which we call TimeSeriesTree. Contracting is enforced by the method timeRemaining. TSF simply builds until it has the required number of trees or the time runs out.

Algorithm 2. buildTSF(A list of n cases length m, $\mathbf{T} = (\mathbf{X}, \mathbf{y})$)

Parameters: the number of trees, k; the minimum interval length, p; the number of
    intervals per tree, r. (default $k \leftarrow 500$, $p \leftarrow 3$, and $r \leftarrow \sqrt{m}$)

1: Let $\mathbf{F} = (\mathbf{F_1} \ldots \mathbf{F_k})$ be the trees in the forest
2: $i \leftarrow 1$
3: **while** $i < k$ **and** timeRemaining() **do**
4:     Let $\mathbf{S}$ be a list of n cases $(\mathbf{s_1} \ldots \mathbf{s_n})$ with $3r$ attributes
5:     **for** $j \leftarrow 1$ to r **do**
6:        $b \leftarrow$ randBetween$(1, m - p)$
7:        $e \leftarrow$ randBetween$(b + p, m)$
8:        **for** $t \leftarrow 1$ to n **do**
9:           $s_{t,3(j-1)+1} \leftarrow$ mean$(\mathbf{x_t}, b, e)$
10:       $s_{t,3(j-1)+2} \leftarrow$ standardDeviation$(\mathbf{x_t}, \mathbf{b}, \mathbf{e})$
11:       $s_{t,3(j-1)+3} \leftarrow$ slope$(\mathbf{x_t}, b, e)$
12:    $\mathbf{F_i}$.buildTimeSeriesTree$(\mathbf{S}, y)$

2.3   Random Interval Spectral Ensemble (RISE)

Like TSF, RISE [15] is a tree based interval ensemble. Unlike TSF, it uses a single interval for each tree, and it uses spectral features rather than summary statistics. RISE was recently updated to be faster and contractable [10]. During the build process, summarised in Algorithm 3, a single random interval is selected for each tree. The first tree is a special case in which the whole series is used (lines 5 and 6). Otherwise, an interval with length that is a power of 2 (line

9) is chosen. The same interval for each series is then transformed using the Fast Fourier Transform (FFT) and Auto Correlation Function (ACF). This is a change on the original RISE which also used the partial autocorrelation function (PACF) and autoregressive (AR) model features. Restriction to the Fourier and ACF coefficients does not decrease accuracy, but makes the algorithm much faster. The power spectrum coefficients are concatenated with the first 100 ACF coefficients to form a new training set. In the `tsml` implementation of RISE the base classifier used is the RandomTree classifier used by random forest (line 13). In the test process class probabilities are assigned as a proportion of base classifier votes.

RISE controls the contract run time by creating an adaptive model of the time to build a single tree (lines 4 and 14). This is important for long series (such as audio), where very large intervals can mean very few trees. Details are in [10].

Algorithm 3. buildRISE(A list of n cases of length m, $\mathbf{T} = (\mathbf{X}, \mathbf{y})$)

Parameters: the number of trees, k; the minimum interval length, p. (default $k \leftarrow$ 500, $p \leftarrow \min(16, m/2)$)

1: Let $\mathbf{F} \leftarrow < \mathbf{F_1} \ldots \mathbf{F_k} >$ be the trees in the forest.
2: $i \leftarrow 1$
3: **while** $i < k$ **and** timeRemaining() **do**
4:     buildAdaptiveTimingModel()
5:     **if** $i = 1$ **then**
6:         $r \leftarrow m$
7:     **else**
8:         $max \leftarrow$ findMaxIntervalLength()
9:         $r \leftarrow$ findPowerOf2Interval(p, max)
10:     $b \leftarrow$ randBetween($1, m - r$)
11:     $\mathbf{T'} \leftarrow$ removeAttributesOutsideOfRange($\mathbf{T}, \mathbf{b}, \mathbf{r}$)
12:     $\mathbf{S} \leftarrow$ getSpectralFeatures($\mathbf{T'}$)
13:     $\mathbf{F_i}$.buildRandomTreeClassifier($\mathbf{S}, y$)
14:     updateAdaptiveModel(r)
15:     $i \leftarrow i + 1$

2.4   Bag of SFA Symbols (BOSS)

Dictionary based classifiers convert real valued time series into a sequence of discrete symbol words, then base classification on these words. Commonly, a sliding window of length w is run across a series. For each window, the real valued series of length w is converted through approximation and discretisation processes into a symbolic string of length l (referred to as a word), which consists of α possible letters. The occurrence in a series of each word from the dictionary defined by l and α is counted and, once the sliding window has completed, the series is transformed into a histogram. Classification is based on the histograms of the words extracted from the series, rather than the raw data. The BOSS [19] ensemble was found to be the most accurate dictionary based classifier in the bake off [3]. Hence, it forms our benchmark for new dictionary based approaches.

Algorithm 4 gives a formal description of the bag forming process of an individual BOSS classifier. Windows may or may not be normalised (lines 6 and 7). Words are created using Symbolic Fourier Approximation (SFA) (lines 8–13). SFA first finds the Fourier transform of the window (line 8), ignoring the first term if normalisation occurs (lines 9–12). It then discretises the first l Fourier terms into α symbols to form a word in the method SFAlookup, using a bespoke supervised discretisation algorithm called Multiple Coefficient Binning (MCB) (line 13). Lines 14–16 implement the process of not counting self similar words: if two consecutive windows produce the same word, the second occurrence is ignored. This is to avoid a slow-changing pattern relative to the window size being over-represented in the resulting histogram.

BOSS uses a non-symmetric distance function in conjunction with a nearest neighbour classifier. Only the words contained in the test instance's histogram (i.e. the word count is above zero) are used in the distance calculation, but it is otherwise the Euclidean distance.

Algorithm 4. baseBOSS(A list of n time series of length m, $\mathbf{T} = (\mathbf{X}, \mathbf{y})$)

Parameters: the word length l, the alphabet size α, the window length w, normalisation parameter z

1: Let $\mathbf{H}$ be a list of n histograms $(\mathbf{h}_1, \ldots, \mathbf{h}_n)$
2: Let $\mathbf{B}$ be a matrix of l by α breakpoints found by MCB
3: **for** $i \leftarrow 1$ to n **do**
4:     **for** $j \leftarrow 1$ to $m - w + 1$ **do**
5:         $\mathbf{s} \leftarrow x_{i,j} \ldots x_{i,j+w-1}$
6:         **if** z **then**
7:             $s \leftarrow$ normalise(s)
8:         $\mathbf{q} \leftarrow$ DFT($\mathbf{s}, l, \alpha, p$) { $\mathbf{q}$ *is a vector of the complex DFT coefficients*}
9:         **if** z **then**
10:            $\mathbf{q}' \leftarrow (q_2 \ldots q_{l/2+1})$
11:        **else**
12:            $\mathbf{q}' \leftarrow (q_1 \ldots q_{l/2})$
13:        $\mathbf{r} \leftarrow$ SFAlookup($\mathbf{q}', \mathbf{B}$)
14:        **if** $\mathbf{r} \neq \mathbf{p}$ **then**
15:            $pos \leftarrow$ index($\mathbf{r}$)
16:            $h_{i,pos} \leftarrow h_{i,pos} + 1$
17:        $\mathbf{p} \leftarrow \mathbf{r}$

The final classifier is an ensemble of individual BOSS classifiers. The original BOSS ensemble built all models over a pre-defined parameter space for w, l, z and α and retained all base classifiers with accuracy of 92% or higher of the best. This introduces instability in memory usage and carries a time overhead. HIVE-COTE 1.0 uses Contractable BOSS (cBOSS) [17] as its dictionary based classifier. cBOSS changes the method used by BOSS to form its ensemble to improve efficiency and allow for a number of usability improvements. cBOSS was shown to be an order of magnitude faster than BOSS on both small and large datasets from the UCR archive while showing no significant difference in accuracy [17]. It randomly samples the parameter space without replacement (line 8), subsamples the data for each base classifier (line 10), and retains a fixed number of base classifiers. An exponential weighting scheme based on train accuracy, such as the one used in HIVE-COTE, is introduced for ensemble members.

Algorithm 5. cBOSS(A list of n cases length m, $\mathbf{T} = (\mathbf{X}, \mathbf{y})$)

Parameters: the maximum number of base classifiers, k; the number of parameter samples, s; the proportion of cases to sample, p. (default $k \leftarrow 50$; $s \leftarrow 250$; $p \leftarrow 0.7$)

1: Let w be window length, l be word length, z be normalise/not normalise and α be alphabet size.
2: Let $\mathbf{B} \leftarrow\; <\mathbf{B_1} \ldots \mathbf{B_k}>$ be a list of k BOSS classifiers
3: Let $\mathbf{W} \leftarrow\; <w_1, \ldots, w_k>$ be a list of classifier weights
4: Let $\mathbf{R}$ be a set of possible BOSS parameter combinations
5: $i \leftarrow 0$, $minAcc \leftarrow \infty$, $idx \leftarrow -1$
6: **while** $i < s$ **and** $|\mathbf{R}| > 0$ **and** timeRemaining() **do**
7: $\quad [l, \alpha, w, z] \leftarrow$ randomSampleParameters($\mathbf{R}$)
8: $\quad \mathbf{R} = \mathbf{R} \setminus \{[l, \alpha, w, z]\}$
9: $\quad \mathbf{T}' \leftarrow$ subsampleData($\mathbf{T}, p$)
10: $\quad$ cls $\leftarrow$ baseBOSS($\mathbf{T}', l, \alpha, w, z$)
11: $\quad acc \leftarrow$ estimateAccuracy($\mathbf{T}', cls$) { *estimate accuracy on train data* }
12: $\quad$ **if** $i < k$ **then**
13: $\quad\quad \mathbf{B_i} \leftarrow$ cls, $w_i \leftarrow acc^4$
14: $\quad\quad$ **if** $acc < minAcc$ **then**
15: $\quad\quad\quad minAcc \leftarrow acc$, $idx \leftarrow i$
16: $\quad$ **else if** $acc > min\_acc$ **then**
17: $\quad\quad \mathbf{B_{idx}} \leftarrow$ cls, $w_{idx} \leftarrow acc^4$
18: $\quad\quad [minAcc, idx] \leftarrow$ findLowestAcc($\mathbf{B}$)
19: $\quad i \leftarrow i + 1$

2.5 Shapelet Transform Classifier (STC)

There are two significant changes to the STC used in HC 1.0. Firstly, it only fully enumerates the shapelet space when there is sufficient time to do so. Secondly, it uses a Rotation Forest classifier [18] rather than a heterogeneous ensemble. The shapelet transform is highly configurable: it can use a range of sampling/search techniques in addition to alternative quality measures. We present the default settings and direct the interested reader to the code. The amount of time for the shapelet search is now a parameter. The algorithm calculates how many possible shapelets there are in a data set, then estimates how many shapelets it can sample from each series. After searching, it updates its timing model using simple reinforcement learning. These operations are encapsulated in operations estimateNumberOfShapelets (line 7) and updateTimingModel (line 11). Shapelets are randomly sampled in the method sampleShapelets (line 8). If the algorithm is allowed more shapelets than the series contains, it evaluates them all. We have experimented with a range of alternative neighbourhood search algorithms, but nothing is much better than random search. Once the shapelets are generated, they are evaluated using information gain (line 9). We use a one vs all evaluation for multi-class problems [5]. Overlapping shapelets are removed in line 10, before the candidates are merged into the overall pool, with the weakest members of the population being deleted. Once the search is complete, the transform is performed (line 13) and the classifier constructed (line 14).

Algorithm 6. STC(A list of n cases length m, $\mathbf{T} = (\mathbf{X}, \mathbf{y})$)

Parameters: the maximum number of shapelets to keep, k; the shapelet search time, t. (default $k \leftarrow 1000$, $t \leftarrow 1$ hour.

1: Let $\mathbf{S}$ be a list of up to k shapelets
2: Let $\mathbf{R}$ be a rotation forest classifier.
3: $i \leftarrow 0$
4: $minIG \leftarrow 0$
5:
6: **while** shapeletTimeRemaining(t) **do**
7:     $p \leftarrow$ estimateNumberOfShapelets(t, m, n)
8:     $\mathbf{S'} \leftarrow$ sampleShapelets($\mathbf{x_i}, p$)
9:     $s \leftarrow$ evaluateShapelets($S', \mathbf{T}$)
10:    $\mathbf{S'} \leftarrow$ removeSelfSimilar(S', s)
11:    updateTimingModel()
12:    $\mathbf{S} \leftarrow$ merge(S, S')
13: $\mathbf{X'} \leftarrow$ shapeletTransform(X, S)
14: $\mathbf{R}$.buildRotationForest($\mathbf{X'}, \mathbf{y}$)

---

3   HIVE-COTE 1.0 Usability

We help maintain two toolkits that include time series classification functionality. sktime[1] is an open source, Python based, sklearn compatible toolkit for time series analysis. sktime is designed to provide a unifying API for a range of time series tasks such as annotation, prediction and forecasting (see [16] for a description of the overarching design of sktime and [1] for an experimental comparison of some of the classification algorithms available). The Java toolkit for time series machine learning, tsml[2], is Weka compatible and is the descendent of the codebase used to perform the bake off. The two toolkits will eventually converge to include all the features described here. Experiments reported in this paper are conducted with tsml, as it has more functionality.

---

[1] https://github.com/alan-turing-institute/sktime.
[2] https://github.com/uea-machine-learning/tsml.

3.1   Java Implementation of HIVE-COTE 1.0 in `tsml`

The HIVE_COTE class is in the package `tsml.classifiers.hybrids` and can
be used as any other Weka classifier. The default configuration is that
described in this paper. The code described here is all available in the class
EX07_HIVE_COTE_Examples with more detail and comments. A basic build is
described in Listing 1.1. It cannot handle missing values, unequal length series
or multivariate problems yet.

```
1   HIVE_COTE hc = new HIVE_COTE();
2   //this setup called in default constructor in April 2020
3   hc.setupHIVE_COTE_1_0();
4   Instances[] trainTest =
5     DatasetLoading.sampleItalyPowerDemand(0);
6   hc.buildClassifier(trainTest[0]);
```

Listing 1.1. A most basic way of using HIVE-COTE 1.0 in tsml

We rarely build the classifier in this way. Instead, we build the components and post process the meta ensemble. This is most easily done using our
Experiments class, which formats the output in a standard way. An example
code snippet is in Listing 1.2. Details on optional input flags not given below
can be found in the code.

```
1   String[] settings=new String[6];
2   //Where to get data
3   settings[0]="-dp=src/main/java/experiments/data/tsc/";
4   //Where to write results
5   settings[1]="-rp=Temp/";
6   //Whether to generate train files or not
7   settings[2]="-gtf=true";
8   //Classifier name: See ClassifierLists for valid options
9   settings[3]="-cn=TSF";
10  //Problem file
11  settings[4]="-dn=Chinatown";
12  //Resample number: 1 gives the default train/test split
13  settings[5]="-f=1";
14  Experiments.ExperimentalArguments expSettings =
15    new Experiments.ExperimentalArguments(settings);
16  Experiments.setupAndRunExperiment(expSettings);
```

Listing 1.2. Using Experiments.java to build a single component.

HIVE_COTE can read component results directly from file using syntax of the
form given in Listing 1.3. It will look in the directory structure created by Experiments. Currently, this file loading method requires all the classifier results to be
present in order to build.

```
1   HIVE_COTE hc=new HIVE_COTE();
2   hc.setBuildIndividualsFromResultsFiles(true);
3   hc.setResultsFileLocationParameters("C:/Temp/", "Chinatown"
      , 0);
```

```
4 String[] components={"TSF","RISE","cBOSS","STC"};
5 hc.setClassifiersNamesForFileRead(components);
```

Listing 1.3. Building HIVE-COTE from existing results files

HIVE_COTE is configurable for different components, threadable (see Listing 1.4) and contractable (see Listing 1.5). In sequential mode, it simply divides the time equally between components. When threaded, it gives the full contract time to each component. It does not yet thread individual components; it is on our development list. You can set the maximum build time for HIVE_COTE if the components all implement the TrainTimeContractable interface.

```
1 HIVE_COTE hc = new HIVE_COTE();
2 EnhancedAbstractClassifier[] classifiers =
3         {new RISE(), new TSF()};
4 String[] names = {"RISE","TSF"};
5 hc.setClassifiers(c, names, null);
6 hc.enableMultiThreading(2);
```

Listing 1.4. Threaded build of HIVE-COTE with bespoke classifiers

```
1 //Ways of setting the contract time
2 HIVE_COTE hc = new HIVE_COTE();
3 //Minute, hour or day limit
4 hc.setMinuteLimit(10);
5 //Specify units
6 hc.setTrainTimeLimit(30, TimeUnit.HOURS);
7 //Or just give it in nanoseconds
8 hc.setTrainTimeLimit(10000000000L);
```

Listing 1.5. Contracting HIVE-COTE for a rom existing results files

Finally, HIVE_COTE is tuneable. Our method of implementing tuned classifiers is to wrap the base classifier in a TunedClassifier object which interacts through the method setOptions. An example of tuning the α parameter is given in Listing 1.6. We advise tuning using results loaded from file. We have found tuning α makes no significant difference. We have not finished evaluating tuning which components to use.

```
1 HIVE_COTE hc=new HIVE_COTE();
2 hc.setBuildIndividualsFromResultsFiles(true);
3 hc.setResultsFileLocationParameters(resultsPath, dataset,
        fold);
4 hc.setClassifiersNamesForFileRead(cls);
5 TunedClassifier tuner=new TunedClassifier();
6 tuner.setClassifier(hc);
7 ParameterSpace pc=new ParameterSpace();
8 double[] alphaVals={1,2,3,4,5,6,7,8,9,10};
9 pc.addParameter("A",alphaVals);
10 tuner.setParameterSpace(pc);
```

Listing 1.6. Tuning HIVE-COTE α parameter from existing results files

3.2   Python Implementation of HIVE-COTE 1.0 in `sktime`

A Python implementation of HIVE-COTE is under development and available in `sktime`. As previously discussed, this version of the algorithm is less mature in terms of its development and the results reported in this paper are from the Java version of the algorithm. The `sktime` implementation is in an alpha state, and will eventually converge on the same functionally as the more developed Java implementation, but currently has a number of limitations (such as building from file and running constituents in parallel). Further, the Python implementations of the constituent classifiers are less efficient than the Java implementations, and as such, HIVE-COTE 1.0 in `sktime` is slower than the `tsml` implementation on equivalent inputs.

The interface and basic usage of HIVE-COTE in `sktime` is very similar to that of `tsml`. The terminology is slightly different however as the `sktime` version uses fit and predict derived from scikit-learn while the Java version uses build and classify from Weka/`tsml`. Notionally the process of constructing and making predictions with HIVE-COTE are equivalent and a simple example of fitting and predicting with HIVE-COTE in `sktime` is given in Listing 1.7. The most up-to-date implementation of HIVE-COTE can be found in the `sktime` toolkit on GitHub under the `hive-cote` branch[3].

```python
def basic_hive_cote(data_dir, dataset_name):
    # using the default constructor for the HIVE-COTE
    class
    hc = HIVE-COTE()

    # loading training data
    train_x, train_y = load_data(
        data_dir + dataset_name + "_TRAIN.ts")

    # building HIVE-COTE 1.0 sequentially
    hc.fit(train_x, train_y)

    # loading testing data
    test_x, test_y = load_data(
        data_dir + dataset_name + "_TEST.ts")

    # predict class values of the test data
    preds = hc.predict(train_x)

    # calculate the test accuracy
    acc = accuracy_score(self.train_y, preds)
```

Listing 1.7. A simple example of using HIVE-COTE 1.0 in `sktime`

---

[3] https://github.com/alan-turing-institute/sktime/blob/hive_cote/sktime/contrib/meta/ensembles.py.

4   Performance

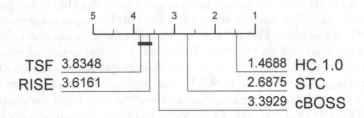

Fig. 2. Critical difference diagram for HIVE-COTE 1.0 and its four components on 112 UCR TSC problems. Full results are available from www.timeseriesclassification/results.php

To measure performance of the new HIVE-COTE, we evaluate each component and the algorithm itself on 112 of the 128 UCR archive datasets. These 112 datasets are all equal length and have no missing values. Figure 2 shows the critical difference diagram for HIVE-COTE 1.0 (HC 1.0) and its four components. This broadly mirrors the performance presented in [15]. We have compared the results of the four components to the original results and found there is no significant difference. Comparison of HIVE-COTE alpha version and 1.0 identify a small, but significant, difference. Removing EE makes HIVE-COTE worse on 48 and better on 33 of the 85 datasets used in the original experiments. The mean reduction in accuracy is 0.6%. The differences in accuracy on specific problems identify those where a distance based approach may be the best. MedicalImages, SonyAIBORobotSurface1, WordSynonyms and Lightning7 were all more than 5% less accurate with EE removed. We are trading this small loss in average accuracy for significant gains in run time and reduction in memory usage. We have explored ways of making EE more efficient, but as yet none of these approaches have met the criteria for maintaining accuracy and providing sufficient speed up. Since HIVE-COTE alpha was proposed, three new algorithms have achieved equivalent accuracy. TS-CHIEF [20] is a tree ensemble that embeds dictionary, spectral and distance based representations. InceptionTime [9] is a deep learning ensemble, combining five homogeneous networks each with random weight initialisations for stability. ROCKET [7] uses a large number (10,000 by default) of randomly parameterised convolution kernels in conjunction with a linear ridge regression classifier. We use the configurations of each classifier described in their respective publications. Figure 3 shows the critical difference diagram for these three classifiers and HC 1.0. There is no significant difference between any of them. The differences between HIVE-COTE and the other three are summarised in Table 1. TS-CHIEF is the most similar to HIVE-COTE, with an average accuracy just 0.25% lower and a high correlation between accuracies. ROCKET has a high variation in performance, as reflected in the high standard

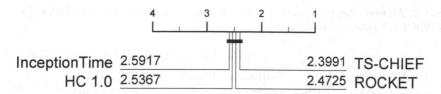

Fig. 3. Critical difference diagram for current state of the art on 109 UCR TSC problems.

deviation of differences and the difference between the mean and median of the differences. It also has the lowest correlation to HC 1.0.

Table 1. Summary of differences in accuracy between HIVE-COTE and the other three algorithms. A negative difference indicates HIVE-COTE is more accurate.

	Mean	Median	Std Dev of differences	Correlation
TS-CHIEF	−0.25%	0.00%	3.801	95.54%
InceptionTime	−0.65%	0.00%	5.82	89.46%
ROCKET	−1.41%	0.05%	8.64	80.28%

Table 2 shows the results for problems with the 10 biggest differences between HIVE-COTE and ROCKET. ROCKET does very poorly on some problems (hence the large average difference in Table 1). InceptionTime also shares this characteristic of occasionally simply failing on a problem. It is not obvious why this happens to both ROCKET and InceptionTime. It may be the result of overfitting. It is worthwhile considering the run time of these algorithms. However, comparisons are made more difficult because of the different software. HIVE-COTE and TS-CHIEF are both built using `tsml`, so are directly comparable. Table 3 summarises the time taken to train the classifiers on 109 UCR problems. Three problems (HandOutlines, NonInvasiveFetalECGThorax1 and NonInvasiveFetalECGThorax2) are omitted because TS-CHIEF did not complete within 7 days (the job limit on our cluster). Of the HC-1.0 components, STC is by far the slowest. This is caused by the classifier, Rotation Forest, not the transform, which is contracted to take at most one hour. The STC design principle is to choose a large number of shapelets (up to 1000) and let the classifier sort out their importance. Rotation forest is on average the best approach for problems with all real valued attributes [2], and we have developed a contracted version that can, if necessary, be used to control the build time (see Listing 1.6). Most of the extra computation required by HIVE-COTE is in forming the estimates of the accuracy on the train data. TS-CHIEF is the slowest algorithm, but it is approximately the same as HIVE-COTE. We do not have reliable

Table 2. Accuracy for ten problems with the biggest difference between HIVE-COTE and ROCKET (five negative and five positive differences).

	TS-CHIEF	ROCKET	HC 1.0	InceptionTime
PigAirwayPressure	96.01%	19.55%	95.77%	92.21%
SemgHandMovementCh2	88.50%	65.26%	88.90%	55.10%
EthanolLevel	60.56%	62.53%	84.90%	87.55%
CinCECGTorso	95.34%	86.41%	99.37%	83.28%
ScreenType	59.42%	60.90%	72.42%	70.56%
FiftyWords	84.27%	82.51%	77.16%	82.68%
ChlorineConcentration	66.08%	79.61%	73.39%	86.36%
MedicalImages	79.91%	80.51%	74.04%	79.63%
WordSynonyms	79.37%	76.44%	69.32%	75.18%
SonyAIBORobotSurface1	88.97%	95.81%	82.63%	95.42%

single-core times for ROCKET, which is run in `sktime`. It is undoubtedly faster than HIVE-COTE and TS-CHIEF though, by at least an order of magnitude.

Table 3. Time in hours to train a classifier for 112 of the UCR problems on a single core.

	TSF	cBOSS	RISE	STC	HC 1.0	TS-CHIEF
Mean	0.13	0.11	0.15	2.30	4.26	5.75
Total	14.15	12.31	16.05	251.12	464.49	626.33

Test time can be a factor for deploying classifiers in time critical situations. Table 4 summarises the time (in minutes) taken to predict the test cases. STC is the slowest component when testing, again caused by the classifier not the transform. TS-CHIEF is the slowest in testing.

Table 4. Time in minutes to make predictions on the test data for 109 of the UCR problems

	TSF	cBOSS	RISE	STC	HC 1.0	TS-CHIEF
Mean	0.01	0.08	0.07	0.62	0.78	4.09
Total	1.42	8.63	7.09	67.50	84.64	445.55

We can also measure maximum memory usage of the classifiers, as this is also often a serious bottleneck for scalability. Table 5 summarises the memory usage of the six `tsml` classifiers. The pattern is similar to that observed with run time. TSF, RISE and cBOSS have a light memory footprint. STC, and hence HIVE-COTE, have a larger memory usage. TS-CHIEF is the most memory hungry, and it seemingly does not scale well in terms of memory. For, example, on the 11 problems where HIVE-COTE required more than 1 GB, TS-CHIEF required approximately four times the memory. The highest memory usage on the 109 problem it could complete was 18 GB on FordA and FordB. It seems highly likely that the memory usage of TS-CHIEF could be improved without loss of accuracy.

Table 5. Memory usage in MB for 109 UCR problems

	TSF	cBOSS	RISE	STC	HC 1.0	TS-CHIEF
Mean	162	255	263	1,464	1,618	2,004
Max	1061	3432	740	3,533	4,426	18,532

5   Conclusions

The purpose of this report is to present a more practical version of HIVE-COTE and compare it to recent advances in the field of TSC. On average, there is no real difference between InceptionTime, TS-CHIEF, ROCKET and HIVE-COTE in terms of accuracy. ROCKET is undoubtedly the fastest, but it is prone to fail badly on the occasional data set. InceptionTime is slow and requires a GPU. It also fails on some data sets. HIVE-COTE and TS-CHIEF are broadly comparable. HIVE-COTE is presently more configurable and controllable.

We hope the results presented here and on the accompanying website serve to act as a baseline comparison for any new research in the field in the future. We have presented HIVE-COTE as is, without improvements that we have been working on. Our next goal is to find a more accurate version that is comparable in run time and memory usage, and to demonstrate scalability by introducing some much larger problems into the UCR archive. We are also developing versions that can handle unequal length series and multivariate TSC problems.

References

1. Bagnall, A., Király, F., Löning, M., Middlehurst, M., Oastler, G.: A tale of two toolkits, report the first: benchmarking time series classification algorithms for correctness and efficiency. ArXiv e-prints. arXiv:1909.05738 (2019)
2. Bagnall, A., Bostrom, A., Cawley, G., Flynn, M., Large, J., Lines, J.: Is rotation forest the best classifier for problems with continuous features? ArXiv e-prints. arXiv:1809.06705 (2018)
3. Bagnall, A., Lines, J., Bostrom, A., Large, J., Keogh, E.: The great time series classification bake off: a review and experimental evaluation of recent algorithmic advances. Data Min. Knowl. Disc. **31**(3), 606–660 (2016). https://doi.org/10.1007/s10618-016-0483-9
4. Bagnall, A., Lines, J., Bostrom, A., Large, J., Keogh, E.: The great time series classification bake off: a review and experimental evaluation of recent algorithmic advances. Data Min. Knowl. Disc. **31**(3), 606–660 (2016). https://doi.org/10.1007/s10618-016-0483-9

5. Bostrom, A., Bagnall, A.: Binary shapelet transform for multiclass time series classification. In: Hameurlain, A., Küng, J., Wagner, R., Madria, S., Hara, T. (eds.) Transactions on Large-Scale Data- and Knowledge-Centered Systems XXXII. LNCS, vol. 10420, pp. 24–46. Springer, Heidelberg (2017). https://doi.org/10.1007/978-3-662-55608-5_2

6. Dau, H., et al.: The UCR time series archive. IEEE/CAA J. Automatica Sinica **6**(6), 1293–1305 (2019)

7. Dempster, A., Petitjean, F., Webb, G.I.: ROCKET: exceptionally fast and accurate time series classification using random convolutional kernels. arXiv preprint arXiv:1910.13051 (2019)

8. Deng, H., Runger, G., Tuv, E., Vladimir, M.: A time series forest for classification and feature extraction. Inf. Sci. **239**, 142–153 (2013)

9. Fawaz, H., et al.: InceptionTime: finding AlexNet for time series classification. ArXiv (2019)

10. Flynn, M., Large, J., Bagnall, T.: The contract random interval spectral ensemble (c-RISE): the effect of contracting a classifier on accuracy. In: Pérez García, H., Sánchez González, L., Castejón Limas, M., Quintián Pardo, H., Corchado Rodríguez, E. (eds.) HAIS 2019. LNCS (LNAI), vol. 11734, pp. 381–392. Springer, Cham (2019). https://doi.org/10.1007/978-3-030-29859-3_33

11. Hills, J., Lines, J., Baranauskas, E., Mapp, J., Bagnall, A.: Classification of time series by shapelet transformation. Data Min. Knowl. Disc. **28**(4), 851–881 (2013). https://doi.org/10.1007/s10618-013-0322-1

12. Large, J., Lines, J., Bagnall, A.: A probabilistic classifier ensemble weighting scheme based on cross validated accuracy estimates. Data Min. Knowl. Disc. **33**(6), 1674–1709 (2019). https://doi.org/10.1007/s10618-019-00638-y

13. Lines, J., Bagnall, A.: Time series classification with ensembles of elastic distance measures. Data Min. Knowl. Disc. **29**(3), 565–592 (2014). https://doi.org/10.1007/s10618-014-0361-2

14. Lines, J., Taylor, S., Bagnall, A.: HIVE-COTE: the hierarchical vote collective of transformation-based ensembles for time series classification. In Proc. 16th IEEE International Conference on Data Mining (2016)

15. Lines, J., Taylor, S., Bagnall, A.: Time series classification with HIVE-COTE: the hierarchical vote collective of transformation-based ensembles. ACM Trans. Knowl. Disc. Data **12**(5), 35 (2018)

16. Löning, M., Bagnall, A., Ganesh, S., Kazakov, V., Lines, J., Király, F.J.: A unified interface for machine learning with time series. ArXiv e-prints. arXiv:1909.07872 (2019)

17. Middlehurst, M., Vickers, W., Bagnall, A.: Scalable dictionary classifiers for time series classification. In: Yin, H., Camacho, D., Tino, P., Tallón-Ballesteros, A.J., Menezes, R., Allmendinger, R. (eds.) IDEAL 2019. LNCS, vol. 11871, pp. 11–19. Springer, Cham (2019). https://doi.org/10.1007/978-3-030-33607-3_2

18. Rodriguez, J., Kuncheva, L., Alonso, C.: Rotation forest: a new classifier ensemble method. IEEE Trans. Pattern Anal. Mach. Intell. **28**(10), 1619–1630 (2006)

19. Schäfer, P.: The BOSS is concerned with time series classification in the presence of noise. Data Min. Knowl. Disc. **29**(6), 1505–1530 (2014). https://doi.org/10.1007/s10618-014-0377-7

20. Shifaz, A., Pelletier, C., Petitjean, F., Webb, G.: TS-CHIEF: a scalable and accurate forest algorithm for time series classification. ArXiv e-prints. arXiv:1906.10329 (2019)

Ordinal Versus Nominal Time Series Classification

David Guijo-Rubio[1]([envelope]) [iD], Pedro Antonio Gutiérrez[1] [iD], Anthony Bagnall[2] [iD], and César Hervás-Martínez[1] [iD]

[1] Department of Computer Sciences, Universidad de Córdoba, Córdoba, Spain
{dguijo,pagutierrez,chervas}@uco.es
[2] School of Computing Sciences, University of East Anglia, Norwich Research Park, Norwich, UK
ajb@uea.ac.uk

Abstract. Time series ordinal classification is one of the less studied problems in time series data mining. This problem consists in classifying time series with labels that show a natural order between them. In this paper, an approach is proposed based on the Shapelet Transform (ST) specifically adapted to ordinal classification. ST consists of two different steps: 1) the shapelet extraction procedure and its evaluation; and 2) the classifier learning using the transformed dataset. In this way, regarding the first step, 3 ordinal shapelet quality measures are proposed to assess the shapelets extracted, and, for the second step, an ordinal classifier is applied once the transformed dataset has been constructed. An empirical evaluation is carried out, considering 7 ordinal datasets from the UEA & UCR Time Series Classification (TSC) repository. The results show that a support vector ordinal classifier applied to the ST using the Pearson's correlation coefficient (R^2) is the combination achieving the best results in terms of two evaluation metrics: accuracy and average mean absolute error. A final comparison against three of the most popular and competitive nominal TSC techniques is performed, demonstrating that ordinal approaches can achieve higher performances even in terms of accuracy.

Keywords: Time series · Ordinal classification · Ordinal regression · Shapelet quality measures

1 Introduction

Time Series Ordinal Classification (TSOC) refers to a prediction problem where the objective is to classify time series with an ordinal label, i.e. the set of labels includes a natural order relationship. In this context, ordinal classification [12] covers those supervised problems where the target variable is discrete and includes a natural order relationship among the labels. Ordinal classification problems can be found in several fields, such as meteorological prediction [10,11], or medical research [19], among others.

© Springer Nature Switzerland AG 2020
V. Lemaire et al. (Eds.): AALTD 2020, LNAI 12588, pp. 19–29, 2020.
https://doi.org/10.1007/978-3-030-65742-0_2

On the other hand, time series consists of data points collected chronologically. In the last years, a countless number of novel approaches in the nominal Time Series Classification (TSC) field have been presented. According to [2], TSC has been tackled from several points of view, depending on the discriminatory features the approach is trying to find. One of these techniques are shapelets [25], phase independent subsequences of the original time series able to differentiate between classes, i.e. a class can be distinguished depending on whether the shapelets could be found in the original time series or not. Further research was done by Hills *et al.* in [13], where the Shapelet Transform (ST) was firstly proposed, in which the best k shapelets (ordered by using a shapelet quality measure) are used to build a transformed dataset in which the attributes are the distances between the shapelets and the original time series. After that, an effective classifier can be applied to the transformed dataset.

Focusing on the proposal of Hills *et al.* in [13], the ST pipeline can be divided into two main steps: 1) the shapelet extraction procedure and 2) the classifier learning using the transformed dataset as input. Regarding the first step, the best k shapelets are selected by using a shapelet quality measure. In order to adapt this approach to the ordinal setting, in this paper, we propose 3 different metrics to measure the ordinal quality of the shapelets, and we compare them against the state-of-the-art Information Gain metric. The second step is adapted by considering an ordinal classifier, instead of using a nominal one, with the objective of exploiting the natural order relationship of the labels. We compare the results obtained against two nominal state-of-the-art techniques.

In this way, the main objectives of this paper are to establish a baseline for TSOC using ST and to demonstrate that, for those ordinal datasets included in the most popular TSC repository, ordinal approaches are able to achieve better performance than standard TSC techniques in terms of accuracy.

2   Background

Time series is a series of data points arranged in time, i.e. the values of the time series are chronological. In a more formal way, the i-th time series object is defined as $\mathbf{T}_i = \{t_{i1}, t_{i2}, \ldots, t_{in}\}$, where n is the length of the time series (note that we only consider equal-length time series). Therefore, a time series dataset is composed of N time series, being defined as $\mathbf{T} = \{\mathbf{T}_1, \mathbf{T}_2, \ldots, \mathbf{T}_N\}$.

In this paper, we are considering ordinal TSC problems: each time series is associated with a label $\mathcal{C}_i \in Y$, where the set of ordinal labels is $Y = \{\mathcal{C}_1, \mathcal{C}_2, \ldots, \mathcal{C}_Q\}$, including $Q > 2$ categories. An order relationship between the labels is found in the problem, i.e. the constraint $C_1 \prec C_2 \prec \ldots \prec C_Q$ should be satisfied.

2.1   Time Series Shapelets

TSC is a very popular field of research in time series data mining [2]. One of the most recent approaches in this field consists in an ensemble including several modules, each one based on a different transformation applied to the original time series dataset, prior to the classification step. One of the first proposals was the shapelet module. A shapelet [25] is a phase independent subsequence of the original time series. The original approach finds all possible shapelets through an enumerative search, which is particularly slow, and then embeds the shapelets in a decision tree without a significant improvement in performance. From this starting point, several approaches have been published in the literature, including [3,9,13], among others. In this paper, we focus on the ST [13], a two-phase approach that uses the extracted shapelets to transform the original dataset, in which the transformed attributes represent the similarity in shape between the original time series and the shapelets obtained, and then applies a classifier to the transformed dataset.

More formally, a shapelet $\mathbf{s}_j = \{s_1, s_2, \ldots, s_l\}$ is a subsequence of a time series $\mathbf{T}_j$, where $l \leq n$ and the subscript j is used to explicitly show that the shapelet $\mathbf{s}$ is a subsequence of time series $\mathbf{T}_j$. The main pipeline for the shapelet extraction procedure consists of three parts [13]: first of all, a set of candidates is randomly generated satisfying several constraints, then, the distance between each shapelet and the original time series is computed to, finally, measure the shapelet quality. The last version of ST [3] proposes new constraints for the shapelet extraction, such as balancing the number of shapelets extracted per class. Moreover, the Euclidean distance is used to measure the similarity between the set of shapelets and the original time series; this distance is computed as the minimum of the distances between the shapelet and all the subsequences with the same length of the shapelet. Finally, the Information Gain (IG) [22] is used to assess the shapelet quality and retain those with higher IG. The formulation is detailed in [13].

In order to consider the natural order between the labels, we propose to consider three different shapelet quality measures. The main idea is to extract shapelets able to reduce the misclassification errors involving more jumps in the ordinal scale:

- Ordinal Fisher (OF) score [20] is an ordinal adaptation of the Fisher score [7]. This measure gives higher penalisation for distant classes in the ordinal scale, therefore, distant classes should be associated with higher distances. It is defined as:

$$OF(\mathbf{s}_j) = \frac{\sum_{k=1}^{Q} \sum_{j=1}^{Q} |\mathcal{O}(\mathcal{C}_k) - \mathcal{O}(\mathcal{C}_j)|(\bar{x}_k - \bar{x}_j)^2}{(Q-1) \sum_{k=1}^{Q} (S_k)^2}, \tag{1}$$

where $\mathcal{O}(\mathcal{C}_q)$ is the position of the category $\mathcal{C}_q$ in the ordinal scale, i.e. $\mathcal{O}(\mathcal{C}_q) = q$, and $\bar{x}_k$ and S_k are the mean and standard deviation of the distances according to the evaluated shapelet $\mathbf{s}_j$ when considering only the time series of the class $\mathcal{C}_k$.
- The Pearson's correlation coefficient (R^2) calculates the correlation between $d_{\mathbf{s}_j, \mathbf{T}_i}$ and c_{y_j, y_i}, $i \in \{1, \ldots, N\}$, where $d_{\mathbf{s}_j, \mathbf{T}_i}$ are the distances from the shapelet $\mathbf{s}_j$ to the original times series $\mathbf{T}_i$, and c_{y_j, y_i} are the differences of

the corresponding class values $c_{y_j, y_i} = |\mathcal{O}(\mathcal{C}_j) - \mathcal{O}(\mathcal{C}_i)|$, where y_j is the class of $\mathbf{T}_j$ (the time series from which the shapelet $\mathbf{s}_j$ is extracted) and y_i is the class of $\mathbf{T}_i$. In this way, R^2 is defined as:

$$R^2(\mathbf{s}) = \sum_{i=1}^{N} \frac{S(d_{\mathbf{s}_j, \mathbf{T}_i}, c_{\mathbf{s}_j, \mathbf{T}_i})}{S_{d_{\mathbf{s}_j, \mathbf{T}_i}} S_{c_{\mathbf{s}_j, \mathbf{T}_i}}}, \tag{2}$$

where $S(\cdot)$ is the covariance of two variables.

– Similarly, the Spearman's correlation coefficient (ρ) computes the correlation between two categorical or continuous variables, following the idea presented for the R^2 quality measure. Therefore, ρ is defined as:

$$\rho(\mathbf{s}) = 1 - \frac{6 \sum_{i=1}^{N} (\mathcal{R}(d_{\mathbf{s}_j, \mathbf{T}_i}) - \mathcal{R}(c_{\mathbf{s}_j, \mathbf{T}_i}))^2}{N(N^2 - 1)}, \tag{3}$$

where $\mathcal{R}(x)$ is the rank of x in the set of all values obtained.

2.2 Ordinal Classification

Once the transformed dataset is constructed (each new attribute j represents the distance between time series i and shapelet j), a classifier is applied to it. One of the main objectives of this paper is to demonstrate that ordinal classifiers can lead to a better performance than nominal ones, given their ability to consider the natural order between the labels. In this way, three different support vector machine techniques have been chosen, using the ORCA framework [21][1]:

– In order to perform comparisons, we first consider nominal Support Vector Classifier (SVC) [14] with two options: one versus one formulation (SVC1V1) and one versus all paradigm (SVC1VA). These two nominal classifiers are very popular in the state-of-the-art, given their accuracy for both binary and nominal multiclass problems.
– On the other hand, an ordinal technique is considered: the Support Vector Ordinal Regression (SVOR) [23] methodology, which is the adaptation of SVC to ordinal classification. Specifically, in this paper we have chosen the SVOR version considering IMplicit constrains (SVORIM) [4]. This approach computes the discriminant parallel hyperplanes for the data and a set of thresholds by imposing implicit constraints in the optimization problem.

In order to assess the performance of ordinal classification problems, there are several metrics that can be considered [5]. In this paper, apart from the accuracy, which is the standard evaluation metric for nominal classification, a specific ordinal evaluation metric should be considered to avoid ignoring order information. In this sense, the misclassification errors are not equally penalised, giving more cost to those misclassified patterns in farther classes. Therefore, we have considered the Correct Classification Rate (CCR), also known as accuracy, which is the global performance of a classifier and the Average Mean Absolute Error ($AMAE$) [1], which measures the ordinal classification errors made for each class.

---

[1] ORCA is available in the repository https://github.com/ayrna/orca.

3 Experimental Results and Discussion

This section exposes the ordinal time series datasets considered, as well as the experimental settings used. Moreover, the results achieved for the three classifiers applied to the four versions of ST using different shapelet quality measures are also shown, along with a comparison of the best ordinal ST approach to the main state-of-the-art algorithms in nominal TSC[2].

3.1 TSOC Datasets

Table 1 shows 7 datasets appropriately selected from the popular UEA & UCR TSC repository[3], given their ordinal nature. All of them belong to the field of bone age estimation [6], except the *EthanolLevel* dataset, which is obtained from the detection of forget spirits using non-intrusive methods [15].

Apart from the main information of the datasets, the Imbalance Ratio (IR) [18] is also included in Table 1. This feature shows whether the distribution of patterns in the datasets is imbalanced, i.e. most of the patterns belongs to a given class (high values for IR). In these cases, trivial classifiers can achieve high values of accuracy.

Table 1. Characteristics of the datasets used in the experiments.

Dataset	#Classes (Q)	#Train	#Test	Length	%IR
DistalPhalanxOutlineAgeGroup	3	400	139	80	1.532
DistalPhalanxTW	6	400	139	80	1.577
EthanolLevel	4	504	500	1751	0.750
MiddlePhalanxOutlineAgeGroup	3	400	154	80	0.881
MiddlePhalanxTW	6	399	154	80	1.276
ProximalPhalanxOutlineAgeGroup	3	400	205	80	0.951
ProximalPhalanxTW	6	400	205	80	2.203

3.2 Experimental Settings

The ST algorithm has been run for one hour during shapelet search. This algorithm has been run with the default values. In the case of ST using IG as shapelet quality measure, an inferior limit of $IG = 0.05$ is used to discard very low-quality shapelets. Furthermore, aiming to reproduce the same behaviour for the remaining shapelet quality measures, the lowest-quality 10% shapelets are also discarded.

---

[2] All the code used in this paper is available in the repository https://github.com/dguijo/TSOC.

[3] http://www.timeseriesclassification.com/.

Regarding the datasets, the standard train and test data splits given in the UEA & UCR TSC repository are used. Moreover, it is worthy of mention that the models are adjusted using only the training set, whereas the test set is only used to evaluate the learned models.

With respect the classifiers, they have been run once, given their deterministic nature. Moreover, their sensitive hyper-parameters have been adjusted using a nested 10-fold cross-validation approach, considering $AMAE$ as the parameter selection criteria, due to the fact that CCR ignores ordinal information. Given that the three classifiers are SVM-based, the same range of values $\{10^{-3}, 10^{-2}, \ldots, 10^3\}$ has been used to adjust both the cost parameter and the kernel width.

Finally, the main code for the ST and for the IG shapelet quality measure was obtained from `sktime` toolkit [17][4].

3.3   Results

In Table 2, the results achieved for the four versions of the ST using different shapelet quality measures are shown. Concretely, the performances of the three classifiers applied to the transforms are presented in terms of CCR and $AMAE$. Furthermore, in order to compare the results in a more global way, we have included the average ranking and the number of datasets in which the respective shapelet quality measure is able to reach to the best performance (#Wins).

As can be seen in Table 2, the ST using R^2 as shapelet quality measure is the one achieving the best results for most of the datasets and classifiers. Specifically, in terms of CCR, the R^2 measure obtains an average ranking of 1.95, followed by ρ (2.36). Regarding number of wins, the ST combined with R^2 reaches to the best results in 11 cases, whereas ST using either ρ or IG ties in 8 cases. On the other hand, in terms of $AMAE$, the ST combined with R^2 also achieves the best results, achieving an average ranking of 1.74 with 11 wins, whereas the second best approach is the standard ST using the IG as shapelet quality measure, with an average ranking of 2.38 and 8 wins. Therefore, it is clear that ST using R^2 as shapelet quality measure achieves the best results without much dependence on the classifier used.

3.4   Comparison Against the State-of-the-Art Algorithms in TSC

In order to establish a comparison against the main state-of-the-art algorithms in TSC, the following three algorithms have been used (which achive the best results up-to-the-knowledge of the authors):

- The Hierarchical Vote Collective of Transformation-based Ensembles (HIVE-COTE) [16] is a meta-ensemble composed of five different modules with several algorithms in each one. These modules rely on the idea of transforming the original dataset prior to classification, such as ST, among others.

---

[4] `sktime` is available in the repository https://github.com/alan-turing-institute/sktime.

Table 2. CCR and $AMAE$ results achieved by the four ST methods (OF, ρ and R^2 are the proposals in this paper).

Classifier	Dataset	CCR				AMAE			
		IG	OF	ρ	R^2	IG	OF	ρ	R^2
SVORIM	DistalPhalanxOutline	**75.54**	74.82	**75.54**	**75.54**	**0.2277**	0.2665	0.2443	**0.2277**
	DistalPhalanxTW	68.35	*69.06*	65.47	**69.78**	**0.4671**	0.5045	0.5264	*0.4822*
	EthanolLevel	**71.40**	46.00	62.00	*62.40*	**0.2938**	0.6067	0.3988	*0.3973*
	MiddlePhalanxOutline	62.99	62.99	**63.64**	**63.64**	**0.5484**	*0.5521*	0.5791	0.5676
	MiddlePhalanxTW	**56.49**	54.55	53.90	**56.49**	1.0137	1.0308	*1.0039*	**0.9851**
	ProximalPhalanxOutline	*86.34*	84.88	*86.34*	**87.32**	*0.1824*	0.2254	0.1978	**0.1744**
	ProximalPhalanxTW	74.63	*76.10*	**79.02**	*76.10*	0.5371	0.4989	*0.4521*	**0.4198**
SVC1V1	DistalPhalanxOutline	**75.54**	74.82	**75.54**	74.82	**0.2277**	0.2334	**0.2277**	0.2334
	DistalPhalanxTW	*69.06*	*69.06*	66.91	**70.50**	0.5600	*0.5046*	0.5440	**0.4614**
	EthanolLevel	**69.00**	48.80	58.20	*61.00*	**0.3301**	0.6795	0.4632	*0.4294*
	MiddlePhalanxOutline	61.04	61.04	*61.69*	**62.34**	0.5827	0.5775	*0.5737*	**0.5636**
	MiddlePhalanxTW	**59.09**	56.49	**59.09**	**59.09**	*0.8785*	0.8962	0.8963	**0.8541**
	ProximalPhalanxOutline	*85.85*	**86.34**	*85.85*	*85.85*	*0.1858*	**0.1820**	0.2016	*0.1858*
	ProximalPhalanxTW	76.59	*78.54*	**80.98**	72.68	0.5104	0.4836	**0.4536**	*0.4569*
SVC1VA	DistalPhalanxOutline	**75.54**	74.10	*74.82*	74.10	**0.2277**	0.2572	*0.2546*	0.2778
	DistalPhalanxTW	66.19	*68.35*	67.63	**69.06**	0.5972	*0.5158*	0.5702	**0.4893**
	EthanolLevel	**67.60**	47.20	56.40	*58.80*	**0.3444**	0.7630	0.5178	*0.4794*
	MiddlePhalanxOutline	*62.34*	61.04	*62.34*	**63.64**	*0.5636*	0.5723	0.5699	**0.5561**
	MiddlePhalanxTW	55.19	49.35	**57.79**	*56.49*	1.0677	1.1290	*0.9689*	**0.9541**
	ProximalPhalanxOutline	*85.85*	85.37	*85.85*	**86.34**	*0.1858*	0.1896	*0.1858*	**0.1820**
	ProximalPhalanxTW	75.61	*76.59*	**80.98**	*76.59*	0.5362	0.4852	**0.3825**	*0.4446*
Average ranking		2.48	3.22	*2.36*	**1.95**	*2.38*	3.24	2.64	**1.74**
#Wins		*8*	1	*8*	**11**	*8*	1	3	**11**

- InceptionTime [8] is an ensemble of deep Convolutional Neural Networks (CNN) models, inspired by the Inception-v4 architecture. In this model, several filters of different lengths are applied simultaneously to an input time series.
- Time Series Combination of Heterogeneous and Integrated Embedding Forest (TS-CHIEF) [24] is an ensemble classifier integrating the most effective embeddings of time series, using tree-structured classifiers.

All these three ensembles are highly competitive in terms of accuracy, although HIVE-COTE is the one achieving the best performance in terms of CCR. However, the main advantages of InceptionTime and TS-CHIEF are their scalability and efficiency.

Table 3 shows the comparison carried out in terms of CCR, given that it is the main goal of TSC. Specifically, the results shown for the ST are those in which the Pearson's correlation coefficient (R^2) is used as the shapelet quality measure, considering different classifiers applied to the transformed dataset: SVC1V1, SVC1VA and SVORIM. Moreover, the results shown for the InceptionTime and

TS-CHIEF algorithms are those presented in the original papers (though we have run the TS-CHIEF algorithm for the *EthanolLevel* dataset, given that it is included in the cited work). For HIVE-COTE, they were obtained using the last version of the algorithm, because it has been recently improved.

Table 3. Comparison in terms of CCR of different classifiers applied to the ST using R^2 as quality measure against the state-of-the-art algorithms in TSC.

	SVC1V1	SVC1VA	SVORIM	HIVE-COTE	InceptionTime	TS-CHIEF
DistalPhalanxOutline	*74.82*	74.10	**75.54**	**75.54**	73.38	74.10
DistalPhalanxTW	**70.50**	69.06	*69.78*	67.63	68.35	68.35
EthanolLevel	61.00	58.80	62.40	*71.40*	**81.40**	52.80
MiddlePhalanxOutline	*62.34*	**63.64**	**63.64**	59.09	55.19	59.09
MiddlePhalanxTW	**59.09**	*56.49*	*56.49*	55.84	51.30	55.85
ProximalPhalanxOutline	85.85	*86.34*	**87.32**	84.39	84.88	84.88
ProximalPhalanxTW	72.68	76.59	76.10	*80.00*	77.56	**81.46**
Average ranking	*3.00*	3.21	**2.36**	3.79	4.43	4.21
#Wins	*2*	1	3	1	1	1

As can be seen in Table 3, SVORIM achieves the best results or the second best in most of the datasets, with 3 wins and an average ranking of 2.36, considerably better than the rest of approaches. SVC1V1 is the second one with an average ranking of 3.00 and 2 wins. The remaining techniques only have 1 win and their average rankings are much worse. Furthermore, all the classifiers applied to the ST combined with R^2 shapelet quality measure (SVC1V1, SVC1VA and SVORIM) achieve a higher performance than state-of-the-art TSC methods.

Some facts must be outlined from the results shown in Table 3: 1) The combination of the ordinal classifier SVORIM with ST R^2 quality measure obtains the best performance in terms of CCR. 2) Nominal classifiers, SVC1V1 and SVC1VA, are taking advantage of the ordinal information induced by the ST combined with R^2 and also obtain competitive results, better than those of the ensemble approaches. 3) HIVE-COTE and TS-CHIEF results are very similar for almost all the datasets, being HIVE-COTE slightly better. 4) InceptionTime is the algorithm obtaining the worse results, because the datasets include short time series. The only exception is *EthanolLevel*, with length equal to 1751, for which InceptionTime is the one obtaining the best performance.

4   Conclusions

This paper presents a novel approach to Time Series Ordinal Classification using the Shapelet Transform (ST). To obtain the k best shapelets for the ST, 3 different ordinal shapelet quality measures are proposed, exploiting the order of labels: Ordinal Fisher (OF), Pearson's correlation coefficient (R^2) and Spearman's correlation coefficient (ρ). These approaches are then compared against Information Gain (IG), which is the one used by the standard ST.

On the other hand, regarding the second step of ST in which a classifier is applied to the transformed data, this paper proposes the use of an ordinal support vector classifier, which is compared against the corresponding nominal versions.

Finally, a comparison against some of the best state-of-the-art techniques in TSC is included: HIVE-COTE, TS-CHIEF and InceptionTime. In this way, the best ordinal approach presented in this paper (ST using R^2 as shapelet quality measure combined with the support vector ordinal classifier) is able to obtain a better accuracy rank than the alternative nominal TSC techniques.

Possible lines of future research are to include the ordinal information of the labels in other points of the ST process and to adapt other modules of the HIVE-COTE meta-ensemble to ordinal classification.

Acknowledgement. This research has been partially supported by the "Ministerio de Economía, Industria y Competitividad" (Ref. TIN2017-85887-C2-1-P) and the "Fondo Europeo de Desarrollo Regional (FEDER) y de la Consejería de Economía, Conocimiento, Empresas y Universidad de la Junta de Andalucía" (Ref. UCO-1261651), Spain. D. Guijo-Rubio's research has been supported by the FPU Predoctoral Program from Spanish Ministry of Education and Science (Grant Ref. FPU16/02128).

References

1. Baccianella, S., Esuli, A., Sebastiani, F.: Evaluation measures for ordinal regression. In: Ninth International Conference on Intelligent Systems Design and Applications, ISDA 2009, pp. 283–287. IEEE (2009). https://doi.org/10.1109/isda.2009.230

2. Bagnall, A., Lines, J., Bostrom, A., Large, J., Keogh, E.: The great time series classification bake off: a review and experimental evaluation of recent algorithmic advances. Data Min. Knowl. Disc. **31**(3), 606–660 (2016). https://doi.org/10.1007/s10618-016-0483-9

3. Bostrom, A., Bagnall, A.: Binary shapelet transform for multiclass time series classification. In: Hameurlain, A., Küng, J., Wagner, R., Madria, S., Hara, T. (eds.) Transactions on Large-Scale Data- and Knowledge-Centered Systems XXXII. LNCS, vol. 10420, pp. 24–46. Springer, Heidelberg (2017). https://doi.org/10.1007/978-3-662-55608-5_2

4. Chu, W., Keerthi, S.S.: Support vector ordinal regression. Neural Comput. **19**(3), 792–815 (2007). https://doi.org/10.1049/cp:19991091

5. Cruz-Ramírez, M., Hervás-Martínez, C., Sánchez-Monedero, J., Gutiérrez, P.A.: Metrics to guide a multi-objective evolutionary algorithm for ordinal classification. Neurocomputing **135**, 21–31 (2014). https://doi.org/10.1109/isda.2011.6121818

6. Davis, L.M., Theobald, B.J., Lines, J., Toms, A., Bagnall, A.: On the segmentation and classification of hand radiographs. Int. J. Neural Syst. **22**(05), 1250020 (2012). https://doi.org/10.1142/s0129065712500207

7. Duda, R.O., Hart, P.E., Stork, D.G.: Pattern Classification. Wiley, Hoboken (2012). https://doi.org/10.1007/s00357-007-0015-9

8. Fawaz, H.I., et al.: InceptionTime: finding AlexNet for time series classification. ArXiv e-prints arXiv:1909.04939, http://arxiv.org/abs/1909.04939 (2019)
9. Grabocka, J., Schilling, N., Wistuba, M., Schmidt-Thieme, L.: Learning time-series shapelets. In: Proceedings of the 20th ACM SIGKDD International Conference on Knowledge Discovery and Data Mining (2014). https://doi.org/10.1145/2623330.2623613
10. Guijo-Rubio, D., et al.: Ordinal regression algorithms for the analysis of convective situations over Madrid-Barajas airport. Atmos. Res. **236**, 104798 (2020). https://doi.org/10.1016/j.atmosres.2019.104798
11. Guijo-Rubio, D., Gutiérrez, P., Casanova-Mateo, C., Sanz-Justo, J., Salcedo-Sanz, S., Hervás-Martínez, C.: Prediction of low-visibility events due to fog using ordinal classification. Atmos. Res. **214**, 64–73 (2018). https://doi.org/10.1016/j.atmosres.2018.07.017
12. Gutiérrez, P.A., Pérez-Ortiz, M., Sánchez-Monedero, J., Fernández-Navarro, F., Hervás-Martínez, C.: Ordinal regression methods: survey and experimental study. IEEE Trans. Knowl. Data Eng. **28**(1), 127–146 (2016). https://doi.org/10.1109/tkde.2015.2457911
13. Hills, J., Lines, J., Baranauskas, E., Mapp, J., Bagnall, A.: Classification of time series by shapelet transformation. Data Min. Knowl. Disc. **28**(4), 851–881 (2013). https://doi.org/10.1007/s10618-013-0322-1
14. Hsu, C.W., Lin, C.J.: A comparison of methods for multiclass support vector machines. IEEE Trans. Neural Netw. **13**(2), 415–425 (2002). https://doi.org/10.1109/72.991427
15. Large, J., Kemsley, E.K., Wellner, N., Goodall, I., Bagnall, A.: Detecting forged alcohol non-invasively through vibrational spectroscopy and machine learning. In: Phung, D., Tseng, V.S., Webb, G.I., Ho, B., Ganji, M., Rashidi, L. (eds.) PAKDD 2018. LNCS (LNAI), vol. 10937, pp. 298–309. Springer, Cham (2018). https://doi.org/10.1007/978-3-319-93034-3_24
16. Lines, J., Taylor, S., Bagnall, A.: Time series classification with HIVE-COTE: The hierarchical vote collective of transformation-based ensembles. ACM Trans. Knowl. Disc. Data **12**(5), 1–35 (2018). https://doi.org/10.1109/icdm.2016.0133
17. Löning, M., Bagnall, A., Ganesh, S., Kazakov, V., Lines, J., Király, F.J.: sktime: a unified interface for machine learning with time series. In: Workshop on Systems for ML at NeurIPS (2019)
18. Pérez-Ortiz, M., Gutiérrez, P.A., Hervás-Martínez, C., Yao, X.: Graph-based approaches for over-sampling in the context of ordinal regression. IEEE Trans. Knowl. Data Eng. **27**(5), 1233–1245 (2014). https://doi.org/10.1109/tkde.2014.2365780
19. Pérez-Ortiz, M., Sáez, A., Sánchez-Monedero, J., Gutiérrez, P.A., Hervás-Martínez, C.: Tackling the ordinal and imbalance nature of a melanoma image classification problem. In: 2016 International Joint Conference on Neural Networks (IJCNN), pp. 2156–2163. IEEE (2016). https://doi.org/10.1109/ijcnn.2016.7727466
20. Pérez-Ortiz, M., Torres-Jiménez, M., Gutiérrez, P.A., Sánchez-Monedero, J., Hervás-Martínez, C.: Fisher score-based feature selection for ordinal classification: a social survey on subjective well-being. In: Martínez-Álvarez, F., Troncoso, A., Quintián, H., Corchado, E. (eds.) HAIS 2016. LNCS (LNAI), vol. 9648, pp. 597–608. Springer, Cham (2016). https://doi.org/10.1007/978-3-319-32034-2_50
21. Sánchez-Monedero, J., Gutiérrez, P.A., Pérez-Ortiz, M.: ORCA: a matlab/octave toolbox for ordinal regression. J. Mach. Learn. Res. **20**(125), 1–5 (2019)

22. Shannon, C.E.: A mathematical theory of communication. ACM SIGMOBILE Mob. Comput. Commun. Rev. **5**(1), 3–55 (2001). https://doi.org/10.1145/584091.584093
23. Shashua, A., Levin, A.: Ranking with large margin principle: two approaches. In: Advances in Neural Information Processing Systems, pp. 961–968 (2003)
24. Shifaz, A., Pelletier, C., Petitjean, F., Webb, G.: TS-CHIEF: a scalable and accurate forest algorithm for time series classification. ArXiv e-prints arXiv:1906.10329 (2019)
25. Ye, L., Keogh, E.: Time series shapelets: a novel technique that allows accurate, interpretable and fast classification. Data Min. Knowl. Disc. **22**(1–2), 149–182 (2011). https://doi.org/10.1007/s10618-010-0179-5

Generalized Chronicles for Temporal Sequence Classification

Yann Dauxais[1] and Thomas Guyet[2]([⊠])

[1] KU Leuven, Celestijnenlaan 200a, Leuven, Belgium
yann.dauxais@kuleuven.be
[2] Institut Agro/IRISA UMR6074, Rennes, France
thomas.guyet@irisa.fr

Abstract. Discriminant chronicle mining (DCM) [6] tackles temporal sequence classification by combining machine learning and chronicle mining algorithms. A chronicle is a set of events related by temporal boundaries on the delay between event occurrences. Such temporal constraints are poorly expressive and discriminant chronicles may lack of accuracy.

This article generalizes discriminant chronicle mining by modeling complex temporal constraints. We present the generalized model and we instantiate different generalized chronicle models. The accuracy of these models are compared with each other on simulated and real datasets.

Keywords: Temporal patterns · Discriminant patterns · Sequence classification

1 Introduction

Temporal sequences, *i.e.*, sequences of timestamped events, are broadly encountered in various applications. They may represent customer purchases, logs of monitoring systems, or patient care pathways and their analysis is highly valuable to support experts 1) to better understand underlying processes and 2) to decide future actions. Face to the large amount of such data, sequence mining techniques have been proposed to extract interesting behaviors. While most sequence mining approaches are dedicated to the extraction of frequent behaviors, few pattern mining approaches deal with discriminant patterns. Discriminant patterns address the task of sequence classification. In a set of labeled sequences, a discriminant pattern associated to a label L occurs more likely in sequence labeled with L than in the other sequences. Discriminant patterns describe the classes of sequences but they can also be used to predict labels of new sequences.

In this work, we assume that temporal information is an important feature to accurately discriminate behaviors. For instance, knowing the delay between two successive visits on a commercial web site may distinguish loyal customers from the others. The sequence of visited pages may be the same, it is the delay between the visit that witnesses the customer loyalty.

© Springer Nature Switzerland AG 2020
V. Lemaire et al. (Eds.): AALTD 2020, LNAI 12588, pp. 30–45, 2020.
https://doi.org/10.1007/978-3-030-65742-0_3

Dauxais et al. [6] introduced chronicles to discriminate temporal behaviors in temporal sequences. A chronicle is a set of events linked by temporal relations imposing numerical bounds on delays between events. We showed that the temporal information captured by chronicles improves the accuracy of sequence labeling.

Mining discriminant chronicles is similar to a regular classification problem. It consists in finding suitable boundaries on the temporal delay to accurately discriminate classes. But, chronicles express very simple boundaries, i.e., delays belonging to an interval.

In this article, we extend the expressiveness of the temporal constraints discovered in chronicles to study the discriminatory power of different types of temporal constraints. The main contribution is the proposal of the generalized discriminant chronicles (GDC). GDC is a meta-model that enables to represent different types of patterns, characterized by their modeling of temporal relations between events. Our framework includes a unified GDC mining procedure inspired by the DCM algorithm [6] and a unified decision procedure to label new sequences. The experiments compare the accuracy of four instances of GDC on simulated and real datasets.

2 Related Works

Sequential patterns have been studied since the early stage of the field of pattern mining [18]. Mabroukeh et al. [13] review the most efficient sequential pattern approaches. All of them are based on the anti-monotonicity property of the pattern support which states that larger patterns occur fewer times in sequences.

In temporal sequences, events are timestamped and our assumption is that the temporal dimension is a key dimension to accurately characterize interesting behaviors. While sequential patterns capture only information about the order of occurrences of events, temporal patterns capture a more expressive temporal information. Different proposals have been made to enrich sequential patterns with more complex temporal information. Mannila et al. proposed episodes [14] as a pattern type which could combine parallel or serial events. Hoeppner et al. [11] introduced Allen's temporal logic to specify the temporal relations between interval events. Two events are not necessarily sequentially ordered, they could "overlap" or "be covered". The temporal relations that are discovered are qualitative. In temporally annotated sequences (TAS) [10] the successive events are constrained by numerical duration extracted by combining a density clustering technique. The chronicle model [5] is at a crossroad between episode and TAS. It is a partial temporal order applied on pattern events constrained by numerical temporal intervals. This pattern model is more general than sequential patterns, TAS and episodes.

Finally, quantitative episodes [15] are tree-based patterns that are graphically similar to chronicles but formally more similar to sets of TAS. Indeed, a quantitative episode represents a set of TAS that are all specifying the same sequential pattern. This set of TAS is represented by a tree rooted on the first event of the sequential pattern for which each path leading to a leaf is a TAS.

Sequence classification has been addressed with statistical approaches such as HMM but also with machine-learning approaches such as recurrent neural networks (LSTM) [12].

Bringmann et al. [3] reviewed "pattern-based classification" that combines pattern mining algorithms and machine learning algorithms to classify structured data, such as sequences. This problem is quite similar to the subgroup discovery task [2]. The main difference between both approaches is that subgroup discovery is meant in a descriptive way whereas pattern-based classification is meant in a predictive way.

The main steps of pattern-based classification are the following (1) a pattern mining step building a vector representation of sequences based on the presence or absence of some extracted patterns; and (2) a machine learning algorithm building a classifier based on the vector representations of labeled sequences. The use of a final classifier makes the results difficult to interpret. For this reason, we focus our interest on the extraction of discriminant patterns, i.e., patterns that can be interpreted by their own as a discriminant behavior.

The proposed approaches are based on interestingness measures different from frequency and capturing the differences between occurrences with subsets of sequences. The most-used measures are growth rate [7] and disproportionality [1]. The BIDE-D algorithm [9] extracts discriminant sequential patterns instead of frequent ones. This technique allows to use a smaller pattern set than the frequent one with a similar accuracy. Recently, the DCM algorithm [6] extended the discriminant sequential pattern mining with chronicles. But temporal constraints of chronicles (i.e., inter-event duration, so-called time gap, within an interval) is maybe too simple to capture complex temporal relationships, and mining patterns with more complex temporal constraints may improve classification accuracy.

3   Discriminant Chronicle Mining

Let $\mathbb{E}$ be a set of event types totally ordered by $\leq_{\mathbb{E}}$. An *event* is a pair (e, t) such that $e \in \mathbb{E}$ and $t \in \mathbb{R}$. A *sequence* is a tuple $\langle SID, \langle (e_1, t_1), (e_2, t_2), \ldots, (e_n, t_n) \rangle, L \rangle$ where SID is the sequence index, $\langle (e_1, t_1), (e_2, t_2), \ldots, (e_n, t_n) \rangle$ a finite sequence of events and $L \in \mathbb{L}$ where $\mathbb{L}$ is a label set. Sequence events are ordered by timestamps and by labels if equality.

Table 1 represents a set of six sequences containing five event types (A, B, C, D and E) and labeled with two different labels $\mathbb{L} = \{+, -\}$. In such case $\leq_{\mathbb{E}}$ is the lexicographic order.

A *chronicle* is a couple $(\mathcal{E}, \mathcal{T})$ such that: $\mathcal{E} = \{\!\{ e_1 \ldots e_n \}\!\}$, $e_i \in \mathbb{E}$ and $e_i \leq_{\mathbb{E}} e_j$ for all $1 \leq i < j \leq n$. $\mathcal{E}$ is a *multiset*, i.e. $\mathcal{E}$ can contain several occurrences of a same event type. $\mathcal{T}$ is a set of *temporal constraints*, i.e. expressions of the form $(e_i, i)[t^-, t^+](e_j, j)$ such that $i, j \in [n]$, $i < j$ and $t^-, t^+ \in \mathbb{R} \cup \{-\infty, +\infty\}\}$. A temporal constraint specifies acceptable delays between the occurrences of the multiset events.

Table 1. Sequences labeled with two classes $\{+, -\}$.

SID	Sequence	Label
s_1	(A, 1), (B, 3), (A, 4), (C, 5), (C, 6), (D, 7)	+
s_2	(B, 2), (D, 4), (A, 5), (C, 7)	+
s_3	(A, 1), (B, 4), (C, 5), (B, 6), (C, 8),(D, 9)	+
s_4	(B, 4), (A, 6), (E, 8), (C, 9)	−
s_5	(B, 1), (A, 3), (C, 4)	−
s_6	(C, 4), (B, 5), (A, 6), (C, 7), (D, 10)	−

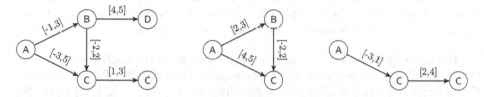

Fig. 1. Examples of three chronicles occurring in Table 1 (detailed in the text). From left to right, the chronicles $\mathcal{C}$, $\mathcal{C}_1$ and $\mathcal{C}_2$.

A chronicle $\mathcal{C} = (\mathcal{E} = \{\!\{e'_1, \ldots, e'_m\}\!\}, \mathcal{T})$ *occurs* in a sequence $s = \langle (e_1, t_1), \ldots, (e_n, t_n) \rangle$, denoted $\mathcal{C} \in s$, iff there exists an injective function $f : [m] \mapsto [n]$ such that 1) $\tilde{s} = \langle (e_{f(1)}, t_{f(1)}), \ldots, (e_{f(m)}, t_{f(m)}) \rangle$ is a subsequence of s, 2) $\forall i, e'_i = e_{f(i)}$ and 3) $\forall i, j, t_{f(j)} - t_{f(i)} \in [t^-, t^+]$ where $e_{f(i)}[t^-, t^+]e_{f(j)} \in \mathcal{T}$. An *occurrence* of $\mathcal{C}$ in s is a list of timestamps $\mathcal{O} = \langle o_1, \ldots, o_m \rangle$ where $\forall i \in [m]$, $o_i = t_{f(i)} \in \mathbb{R}$.

The *support* of a chronicle $\mathcal{C}$ in a set of sequences $\mathcal{S}$ is the number of sequences in which $\mathcal{C}$ occurs:

$$supp(\mathcal{C}, \mathcal{S}) = |\{s \in \mathcal{S} \mid \mathcal{C} \in s\}|.$$

Figure 1 illustrates three chronicles represented by graphs. Chronicle $\mathcal{C} = (\mathcal{E}, \mathcal{T})$ where $\mathcal{E} = \{\!\{A, B, C, C, D\}\!\}$ and $\mathcal{T} = \{(A, 1)[-1, 3](B, 2), (A, 1)[-3, 5](C, 3), (B, 2)[-2, 2](C, 3), (B, 2)[4, 5](D, 5), (C, 3)[1, 3](C, 4)\}$ is illustrated at the top left. Chronicle $\mathcal{C}$ (see Fig. 1 on the left), occurs in sequences s_1, s_3 and s_6 of Table 1. We notice there are two occurrences of $\mathcal{C}$ in sequence s_1. Nonetheless, its support is $supp(\mathcal{C}, \mathcal{S}) = 3$. The two other chronicles, denoted $\mathcal{C}_1$ and $\mathcal{C}_2$, occur respectively in sequences s_1 and s_3; and in sequence s_6. Their supports are $supp(\mathcal{C}_1, \mathcal{S}) = 2$ and $supp(\mathcal{C}_2, \mathcal{S}) = 1$.

Frequent chronicle mining consists in extracting all chronicles $\mathcal{C}$ in a dataset $\mathcal{S}$ such that $supp(\mathcal{C}, \mathcal{S}) \geq \sigma_{min}$ [5]. The *DCM* algorithm extracts *discriminant chronicle* [6]. A discriminant chronicle occurs at least g_{min} times more in the set of positive sequences, *i.e.* sequences labeled with +, than in the set of negative sequences (labeled with −). Then, it can be represented as a classification rule $\mathcal{C} \Rightarrow +$ specifying that sequences in which $\mathcal{C}$ occurs more likely belongs to class $+$. In this mining task, the user has to specify two thresholds: the minimum frequency threshold σ_{min} and the minimal growth rate $g_{min} \geq 1$.

4   Generalized Discriminant Chronicles (GDC)

We now introduce the generalized discriminant chronicle (GDC) meta-model. The GDC meta-model defines an abstract pattern model of temporal behaviors. Next section instantiates different concrete approaches within a unified framework of generalized discriminant chronicle, i.e. a GDC model and a mining algorithm (see Sect. 4.2).

Let $\mathbb{L}$ be a set of labels and $\mathbb{E}$ be a set of event types, a *generalized discriminant chronicle* (GDC) is a couple $(\mathcal{E}, \mu)$, where $\mathcal{E}$ is a multiset of event types and $\mu : \mathbb{R}^{|\mathcal{E}|} \mapsto [0,1]^{|\mathbb{L}|}$ is an *occurrence assessment function*.

The *occurrence assessment function* intuitively gives the confidence measure that a multiset witnesses each label. For some occurrence $\mathcal{O} \in \mathbb{R}^{|\mathcal{E}|}$ of multiset $\mathcal{E}$ in a sequence, $\mu(\mathcal{O}) = [p_1, p_2, \ldots, p_{|\mathbb{L}|}]$ where $\forall i$, $\mu^i(\mathcal{O}) = p_i \in [0,1]$ gives the confidence measure that $\mathcal{O}$ belongs to the i-th class. In case it sum to 1, this vector can be interpreted as a probability distribution.

GDC generalizes the previous definition of chronicles in two manners: 1) the *occurrence assessment function* is a generalization of the temporal constraints, 2) the *weighted vector of decisions* $[0,1]^{|\mathbb{L}|}$ is the generalization of the association of a chronicle to a label ($\mathcal{C} \Rightarrow L, L \in \mathbb{L}$).

In particular, it is possible to encode the discriminant chronicle $(\mathcal{E}, \mathcal{T}) \Rightarrow L_l$, where $L_l \in \mathbb{L}$ is the l-th sequence class, as a GDC using $\mu_{\mathcal{T}}$ defined such that for some occurrence $\mathcal{O} = \{o_i\}_{i \in [|\mathcal{E}|]}$ of $\mathcal{E}$:

$$\mu_{\mathcal{T}}(\mathcal{O}) = \begin{cases} \mathbb{1}_l \text{ if } \forall e_i[a,b]e_j \in \mathcal{T}, a \leq o_j - o_i \leq b \\ \mathbf{0} \text{ otherwise} \end{cases}$$

where $\mathbb{1}_l$ is a vector of zeros except at position l (value 1), and $\mathbf{0}$ is a vector of zeros. The size of these two vectors is $|\mathbb{L}|$.

4.1   Taking Decisions with Generalized Discriminant Chronicles

This section describes how generalized discriminant chronicles are used to automatically classify new sequences. Let $C = (\mathcal{E}, \mu)$ be a GDC and s be a sequence to classify such that there exists at least one occurrence $\mathcal{O} \in \mathbb{R}^{|\mathcal{E}|}$ of multiset $\mathcal{E}$. Then, decision vector is given by $\mu(\mathcal{O})$. But the multiset $\mathcal{E}$ may occur several times in s. All decisions have to be combined and the final classification decision for sequence s, denoted $d_C(s) \in \mathbb{L}$, is the class label with the largest confidence value:

$$d_C(s) = \operatorname*{argmax}_{l \in \mathbb{L}} \left(\max_{\mathcal{O}} \mu^l(\mathcal{O}) \right) \tag{1}$$

where $\max_{\mathcal{O}} \mu^l(\mathcal{O})$ denotes the maximum confidence value obtained for label $l \in \mathbb{L}$ for all occurrences $\mathcal{O}$ of the multiset. The function $d_C(s)$ enables to use a GDC as a decision rule. It decides which class a sequence belongs to.

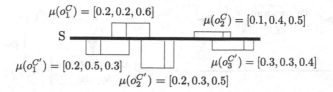

$\mu(o_1^C) = [0.2, 0.2, 0.6]$

$\mu(o_2^C) = [0.1, 0.4, 0.5]$

$\mu(o_1^{C'}) = [0.2, 0.5, 0.3]$ $\mu(o_3^{C'}) = [0.3, 0.3, 0.4]$

$\mu(o_2^{C'}) = [0.2, 0.3, 0.5]$

Fig. 2. Examples of multiple occurrences of two chronicles C and C' to be combined to make the final decision. A "rake" illustrates item positions in the sequence of one occurrence of the multiset.

The class label can also be decided from a set of chronicles $\mathcal{C} = \{C_i\}_{1 \leq i \leq n}$. In this case, each chronicle yields its own decision, and they are merged into a final decision. The decision procedure we propose, denoted $d_{\mathcal{C}}(s)$ – with a collection of chronicles as subscript, is a linear combination of the number of occurrences of a chronicle C_i in s labeled with $l_j \in \mathbb{L}$, more formally:

$$d_{\mathcal{C}}(s) = \operatorname*{argmax}_{j \in \mathbb{L}} \left(\sum_{i=1}^{n} \alpha_{i,j} \nu_{C_i}^j(s) + \beta_j \right) \tag{2}$$

where $\alpha_{i,j} \in \mathbb{R}$ and $\beta_{i,j} \in \mathbb{R}$ are parameters, and $\nu_{C_i}^j(s)$ is the number of occurrences of chronicle C_i in sequence s that suggests classifying the sequence in class j (*i.e.* $d_{C_i}(s) = j$):

$$\nu_{C_i}^j(s) = \left| \left\{ \mathcal{O} \subset \mathbb{R}^{|\mathcal{E}|} \,\middle|\, \operatorname*{argmax}_{l \in \mathbb{L}}(\mu^l(\mathcal{O})) = l_j \right\} \right| \tag{3}$$

Figure 2 illustrates a sequence s classified with a set of two chronicles C and C', and $|\mathbb{L}| = 3$. Chronicle C occurs twice in s and C' occurs thrice, $\mathcal{O} = \{o_1^C, o_2^C, o_1^{C'}, o_2^{C'}, o_3^{C'}\}$. The figure illustrates respective decision vectors.

In this case, $\nu_C = [0, 0, 2]$ because the majority class in the two occurrences of chronicle C is the third one, and $\nu_{C'} = [0, 1, 2]$. Assuming $\beta_j = 0$ and $\alpha_{i,j} = 1$, $\forall i, j$, then the predicted class is $d_{\mathcal{C}}(s) = 3$ because $\sum_{i=1}^{n} \alpha_{i,3} \nu_{C_i}^3(s) + \beta_3 = 4$ is the largest predicted value among possible classes.

The intuition is that the contribution to the final decision of chronicle C_i is more important if this chronicle appears several times in the sequence. Combining the numbers of occurrences is preferred to the combination of confidence measures to prevent from bias due to chronicles with low recall (*i.e.* poorly informative) but with possible high confidence.

In practice, the $\alpha_{i,j}$ and β_j parameters are not set up manually but learned from data as explain in next section.

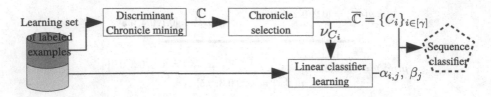

Fig. 3. Overall procedure of sequence classifier learning.

4.2   Learning Generalized Discriminant Chronicles Classifiers

The overall procedure dedicated to learn the sequence classifier is given in Fig. 3. This procedure extracts both a set of γ discriminant chronicles, denoted $\overline{\mathbb{C}}$, and parameters values of decision function (see Eq. 2). First of all, the learning dataset is split into two separated bunches of sequences.

One dataset is used to extract a set of discriminant chronicles $\mathbb{C} = \{C_i\}$. A subset of the γ most discriminant chronicles, denoted $\overline{\mathbb{C}}$, is selected from $\mathbb{C}$. According to BIDE-D [9], reducing the set of chronicles prevents from overfitting. The second dataset is used to learn the decision procedure. In case of a dataset with two classes ($\mathbb{L} = \{+, -\}$), Eq. 2 can be seen as a linear classification problem. Then, a linear-SVM classifier learns the $\alpha_{i,j}$ and β_j parameters. In practice, a linear-SVM classifier is also used for a multi-class setting parameters and its model serves as decision function that takes the final classification decision.

We now come back to the mining of generalized discriminant chronicles. This algorithm is based on the original DCM algorithm [6]. Algorithm 1 gives the general principle of GDC mining from a dataset of labeled sequences $\mathcal{S}$. The two main parameters are σ_{min}, a minimal support threshold used to prevent from generating too much poorly-representative chronicles and g_{min}, a minimal growth rate threshold. The overall principle from learning multiple-class chronicles is the one class against all. For some class L, the minimal growth rate g_{min} indicates that a GDC occurs at least g_{min} times more in sequences of class L than in all other sequences.

Algorithm 1 extracts GDC for each class L in two main steps. It firstly extracts $\mathbb{M}$, the set of frequent multisets in the sequences of class L. Then, EXTRACTDTC learns a μ function from the list of occurrences of a frequent multiset. There is a unique μ per multiset. Its principle is first to build a time-gap table [19] from all occurrences of a multiset and, second, to learn a temporal model from the time-gap table. Each time-gap occurrence is labeled by the label of its sequence and any standard machine learning algorithm can learn the μ function.

The DCM algorithm [6] is based on rule induction (*e.g.* *Ripper* [4]) to learn temporal constraints of a chronicle ($\mathcal{T}$). It is a specific case of a $\mu_{\mathcal{T}}$ function that fits the requirements of the original model of chronicle. Next section introduces alternative classes of occurrence assessment functions.

Algorithm 1. Generalized discriminant chronicle mining

Require: $\mathcal{S}$: labeled sequence sets, $\mathbb{L}$: set of labels, σ_{min}: minimal support threshold, g_{min}: minimal growth threshold
1: $\mathbb{C} \leftarrow \emptyset$ ▷ $\mathbb{C}$ is the discriminant chronicle set
2: **for all** $L \in \mathbb{L}$ **do**
3:     $\mathbb{M} \leftarrow \text{ExtractMultiSet}(\mathcal{S}^L, \sigma_{min})$
4:     **for all** $ms \in \mathbb{M}$ **do**
5:         **for all** $\mu \in \text{ExtractDTC}(\mathcal{S}, L, ms, g_{min}, \sigma_{min})$ **do**
6:             $\mathbb{C} \leftarrow \mathbb{C} \cup \{(ms, \mu)\}$ ▷ Add a new GDC
7: **return** $\mathbb{C}$

---

5   Examples of GDC Instances

This section illustrates several types of patterns that can be represented by GDC: discriminant sequential patterns, discriminant episodes, SVM-DC and DT-DC. The first two types of patterns illustrate the ability of GDC to model existing patterns (less expressive than the original discriminant chronicles) and the last two models illustrate meaningful generalizations of temporal constraints. In the remaining of this section, we briefly present each of these models as instances of the GDC.

Discriminant episodes and sequences An episode is a set of events ordered temporally by a partial order $\leq_{\mathcal{E}}$. If $\leq_{\mathcal{E}}$ is a total order, the episode is a sequential pattern. Such classical temporal patterns have been used for mining discriminant behaviors respectively by Fabrègue et al. [8] and by Fradkin et al. [9]. A discriminant episode is an episode associated to a label $L \in \mathbb{L}$. Such discriminant patterns could be represented by a GDC model instance by the following occurrence assessment function:

$$\mu_{\mathcal{T}}(o) = \begin{cases} \mathbb{1}_L & \text{if } \forall(i,j),\ i \leq_{\mathcal{E}} j \Rightarrow o_i \leq o_j \\ 0 & \text{otherwise} \end{cases} \tag{4}$$

For example, a multiset $\mathcal{E} = \{\!\{A, B, C\}\!\}$ ordered by $\leq_{\mathcal{E}}$ such that $B \leq_{\mathcal{E}} A$ and $B \leq_{\mathcal{E}} C$ specifies an episode representing sequences where B occurs before events A and C, no matter the order between A and C. While associated to a label, it becomes a discriminant episode. Expressed with chronicle temporal constraints, we have $\mathcal{T} = \{(A, 1)[-\infty, 0](B, 2), (A, 1)[-\infty, \infty](C, 3), (B, 2)[0, \infty](C, 3)\}$.

Decision Tree Discriminant chronicles (DT-DC). A discriminant chronicle is characterized by temporal constraints on the time gaps ($\mathcal{T}$). A constraint (e, i) $[t^-, t^+](e', j)$ enforces the time gap δ between some occurrences of events e and e' to belong to the interval $[t^-, t^+]$. But chronicle does not allow disjunctions of constraints. For instance, it is not possible to specify that δ may belong to $[t^-, t^+] \cup [t'^-, t'^+]$.

The DT-DC model replaces the conjunctive rule learning algorithm by a decision tree, such as C4.5 [17]. For example, let's consider a dataset

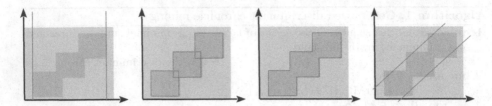

Fig. 4. Illustration of temporal discrimination power of the different instances of GDC. Planes (x, y) represent a pair of temporal constraints for some sequences $(A, t_0)(B, t_0 + x)(C, t_0 + x + y)$. Positive sequences are those with (x, y) values in the green region, negative sequences have (x, y) values in the red region. The bold-green lines represent the separation boundaries learned by a GDC depending on the type of occurrence assessment function, μ. From left to right: discriminant episodes (temporal constraints with shape $[0, +\infty]$), chronicles (three chronicles with temporal constraints represented by rectangles), DT-DC (a single shape combining several rectangles), linear-SVM (a single chronicle, with generalized linear boundaries).

of positive sequences matching temporal constraints $(A, 1)[2, 3](B, 2)$ and $(A, 1)[7, 9](B, 2)$ and a dataset of negative sequences matching the temporal constraint $(A, 1)[2, 9](B, 2)$. In this case, two chronicles would be discriminant (one per interval, $[2, 3]$ and $[7, 9]$). On the opposite, a single DT-DC will capture the disjunction of intervals in the same model. The expected benefit of this model is a better generalization power.

SVM Discriminant chronicles (SVM-DC). SVM Discriminant chronicles illustrate the case of a complex learnable occurrence assessment function μ, *i.e.* a μ modeled by a multi-class *SVM* classifier. Compared to the previous types of patterns, SVM-DC is not limited to linear boundaries to separate examples (time gaps of multiset occurrences) and is a good candidate for yielding accurate patterns.

It is worth noticing that any machine learning model yields a new type of discriminant temporal patterns based on the GDC. The above GDC instances show the potential variety of temporal constraints that GDC can model. Figure 4 illustrates the shape of boundaries defined by occurrence assessment functions of a chronicle learned from a synthetic dataset.

6   Experiments

In this part, we compare different results in pattern-based classification using discriminant episodes, discriminant chronicles, DT-DC and SVM-DC. The goal of these experiments is to highlight the impact of the GDC model choice on the accuracy of decision functions presented in Sect. 4.1: $d_C(s)$ and $d_{\mathcal{C}}(s)$. In the experiments we analyze the classification power of individual GDC (*i.e.* $d_C(s)$) and of a set of GDC (*i.e.* $d_{\mathcal{C}}(s)$).

The DT-DC and SVM-DC mining algorithms are implemented in Python using scikit-learn library [16]. The algorithm dedicated to discriminant chronicle mining is implemented in C++.[1]

6.1 Experimental Setup

The different experiments compare mean accuracy of different GDC models obtains by cross-validation on synthetic and real datasets.

A 5-cross-validation is performed on each dataset for the parameters σ_{min} and g_{min} of the mining step described in Sect. 3 and the parameter γ described in Sect. 4.2. The domains used for σ_{min}, g_{min} and γ are respectively $\{0.2, 0.3, 0.4, 0.5, 0.6\}$, $\{1.4, 1.6, 1.8, 2, 3\}$ and $\{90, +\infty\}$. $\gamma = +\infty$ means that all discriminant chronicles are kept. To improve the computation time, a fourth parameter is introduced for the mining step: the maximal size of extracted chronicles max_size. This parameter constrains the maximal number of events that a GDC, $i.e.$ its multiset, can contain. The domain of this parameter is $\{3, 4, 5, 6\}$.

The real datasets are the UCI datasets presented in the BIDE-D experiments [9]: $asl\text{-}bu$, $asl\text{-}gt$ and $blocks$. These datasets are part of the standard benchmark for pattern-based classification approaches.

We generated two collections of synthetic datasets:

- A first collection of datasets is based on the principle illustrated by Fig. 4. Random sequences $\langle(A, 0)(B, x)(C, x + y)\rangle$ have been generated: the event A occurs at time 0 in each sequence and the time gaps between A and B and between B and C are randomly generated in the interval $[0, 15]$. The label of the sequence is generated depending on the temporal constraints. According to Fig. 4, positive examples having time gaps located in one of the three green squares. Coordinates of the square corners are $(1, 1), (6, 6); (5, 5), (10, 10)$ and $(9, 9), (14, 14)$. Each dataset contains 150 positive and 150 negative sequences.
- A second collection of datasets is based on random sequences with shape $\langle(A, t_A)(B, t_A + x)(C, t_C)(D, t_C + k \times x)\rangle$ where $x \in [1, 9]$, $t_A = 15$ and $t_C \in [1, 29]$. The two sequence classes are distinguished by the k factor: $k = 2$ for positive sequences while $k = 1$ for negative ones. Each dataset contains 100 positive and 100 negative sequences.

To ease the comparison between DT-DC and discriminant chronicles as individual patterns, we choose to use each node of the extracted trees as discriminant temporal constraint. This prevents from comparing the classification power of the decision-tree algorithm and the rule learning algorithm ($Ripper$). Furthermore, decision trees produce more discriminant chronicles than $Ripper$ because tree nodes are more redundant.

---

[1] All software sources and datasets are available at https://gitlab.inria.fr/ydauxais/ GDC-PBC.

Table 2. The five most accurate parameter sets for discriminant chronicles on the first synthetic dataset. The *number* attribute is the total number of chronicles extracted in the 5 runs.

σ_{min}	g_{min}	Accuracy	Support	Number
0.2	1.6	0.92($\pm$0.04)	33.15($\pm$6.95)	5.2($\pm$0.40)
0.2	2	0.88($\pm$0.06)	36.77($\pm$7.79)	4.4($\pm$0.49)
0.2	1.8	0.86($\pm$0.05)	35.41($\pm$8.20)	4.4($\pm$0.80)
0.2	3	0.85($\pm$0.03)	36.57($\pm$7.55)	4.2($\pm$0.40)
0.2	1.4	0.83($\pm$0.04)	36.67($\pm$7.04)	4.8($\pm$0.75)

6.2 Results

Let us first present results obtained by the GDC instances on the first synthetic datasets. For this experiment, we only consider the extracted chronicles with multiset $\{\!\{A, B, C\}\!\}$. Thus, only one DT-DC and one SVM-DC are extracted for each run, but the number of discriminant chronicles depends on the setting (see Table 2). The three squares defining the positive occurrences may be represented by three discriminant chronicles but it is more difficult to represent the negative occurrences because some of them are not included in a frequent rectangle containing only negative occurrences.

The unique DT-DC represents almost perfectly the discriminant behavior used to generate the dataset with a mean accuracy of 0.99($\pm$0.02). This result was expected because of the dataset structure (squares with boundaries orthogonal to the axis) fits the discrimination capabilities of decision trees.

No SVM-DC are extracted for the default parameters of g_{min}. Our explanation is that concavities in the shape containing positive occurrences disadvantage linear SVM. Relaxing the constraint of g_{min}, SVM-DC reaches a mean accuracy of 0.48($\pm$0.05). This shows experimentally that DT-DC can be more accurate than SVM-DC for some datasets. No discriminant episodes are extracted from these datasets. It was expected from their design as each sequence only contains the items A, B and C, and always in the same order. Some discriminant episodes could be extracted from the negative sequences like the one representing A and B occurring at the same time but these patterns are rare and thus they are not extracted using the defined parameters. It is an example of the need to generalize such a simple model in order to catch more complex behaviors.

We compared these results with the discriminant chronicles obtained using DCM. Among the discriminant chronicles extracted using DCM, discriminating positive occurrences from negative ones generates three perfectly discriminant chronicles representing the three squares used to generate the data with parameters $\sigma_{min} = 0.2$, $g_{min} = 3$ and considering only the multiset $\{\!\{A, B, C\}\!\}$. Discriminating the negative occurrences from the positive ones with DCM generates two perfectly discriminant chronicles representing the largest rectangles of negative occurrences on the top left and on the bottom right of the Fig. 4. Then, the mean accuracy of discriminant chronicles is 1 and, contrary to DT-DC, some

Table 3. Five most accurate parameter sets for regular discriminant chronicles on the second synthetic dataset. The attribute *number* is the total number of chronicles extracted in the 5 runs.

σ_{min}	g_{min}	Accuracy	Support	Number
0.3	3.0	1	10.5($\pm$3.46)	16
0.2	1.8	0.95($\pm$0.13)	8.19($\pm$3.64)	37
0.2	1.6	0.90($\pm$0.17)	12.9($\pm$8.86)	31
0.6	1.6	0.79($\pm$0.17)	18.8($\pm$10.79)	11
0.6	1.4	0.72($\pm$0.17)	20.4($\pm$10.48)	10

negative occurrences are not covered by these chronicles. This perfect accuracy is correlated to the strategy of *Ripper* that does not reuse covered occurrences to build a new temporal constraint. The remaining occurrences are so considered too few to be used for building a new constraint. The accuracy is better due to the partial coverage of the dataset made by discriminant chronicles.

We present the same experiment on the second collection of datasets. These datasets are generated to favor the SVM-DC model with boundaries that correlates linearly the time gaps. Again, we only considered the extracted chronicles with multiset $\{A, B, C, D\}$. For the simplest dataset, the single extracted SVM-DC obtained the accuracy of 1 for the 5 runs. The extracted DT-DC obtained an mean accuracy of 0.99($\pm$0.02). Thus, SVM-DC accuracy is not better than DT-DC, but, DT-DC builds a very large decision tree that overfits the boundaries, which is not suitable in real applications.

Table 3 shows the results of regular discriminant chronicles. We observe that the most accurate parameter sets extract chronicles with a small support and the least accurate parameter sets extract chronicles with a bigger support. We do not present results for discriminant episodes because not any discriminant episodes are extracted. This shows the limit of a too simple model.

Let us now present the results obtained by the three GDC models as individual patterns and as pattern sets on real datasets. An overview of the classification power of the individual patterns of DT-DC, discriminant chronicles and discriminant episodes is given by Table 4. It shows that DT-DC patterns are individually less accurate than discriminant chronicles obtained from the same decision trees. Discriminant episodes are also individually more discriminant than discriminant chronicles. The intuition behind these results is that decision trees overfit more the datasets than temporal constraints or sequential orders. Temporal constraints and sequential orders gather only dense sets of occurrences, represented as squares on Fig. 4, but decision trees generalize examples and gather too dissimilar occurrences.

Conversely to the accuracy, the mean support is higher for DT-DC than for discriminant chronicles and discriminant episodes. Each DT-DC covers more examples than the two other types of patterns. Furthermore, the coverage of

Table 4. Comparison table of DT-DC, discriminant chronicles and discriminant episodes for the 5 best parameter sets in terms of mean accuracy on *asl-bu*.

	σ_{min}	g_{min}	max_size	Accuracy	Support
Discriminant episodes	0.3	2.0	3	0.92($\pm$0.14)	5.62($\pm$8.60)
	0.3	1.8	4	0.86($\pm$0.19)	5.24($\pm$7.65)
	0.6	3.0	3	0.86($\pm$0.20)	2.96($\pm$5.54)
	0.6	1.8	4	0.83($\pm$0.18)	11.41($\pm$9.71)
	0.3	1.6	3	0.83($\pm$0.22)	4.41($\pm$5.94)
Discriminant chronicles	0.3	2.0	3	0.64($\pm$0.37)	3.16($\pm$2.91)
	0.3	1.8	4	0.60($\pm$0.40)	2.69($\pm$2.68)
	0.6	3.0	3	0.59($\pm$0.33)	5.74($\pm$7.01)
	0.3	1.6	3	0.58($\pm$0.38)	3.67($\pm$3.28)
	0.6	1.8	4	0.57($\pm$0.34)	3.83($\pm$3.37)
DT-DC	0.5	2.0	5	0.38($\pm$0.28)	8.12($\pm$6.57)
	0.6	3.0	5	0.37($\pm$0.25)	8.61($\pm$6.34)
	0.2	1.6	3	0.36($\pm$0.34)	5.99($\pm$5.68)
	0.5	1.8	3	0.34($\pm$0.21)	13.2($\pm$8.19)
	0.5	1.8	4	0.33($\pm$0.23)	10.0($\pm$7.40)

discriminant episode is worse than DT-DC due to their poor expressiveness of temporal behaviors.

These accuracy results are extreme because we did not use the decision tree parameter constraining a leaf to have a minimal support. For example, if this parameter is set to f_{min}, a decision tree can be seen as a set of discriminant chronicles for a unique multiset.

To illustrate the importance of this parameter, we can compare the two accuracy distributions. Figure 5 at top left shows the accuracy distribution of DT-DC for the most accurate parameter set. The distribution at bottom right is the accuracy distribution of discriminant chronicles for the most accurate parameter set. The distribution at bottom left is the accuracy distribution of DT-DC for the best discriminant chronicle parameter set with a mean accuracy of 0.23($\pm$0.25). The distribution at top right is the accuracy distribution of discriminant chronicles for the best DT-DC parameter set with a mean accuracy of 0.32($\pm$0.39).

We first notice that the accuracy distributions of discriminant chronicles and DT-DC are almost similar. These histograms show three main peaks: patterns that obtained the accuracy of 0, 0.5 and 1. It makes sense considering that both types of patterns were extracted by the same algorithm. The differences between the mean accuracy are mainly in the proportion of patterns with accuracy equals to 0 and equals to 1. Proportionally to the number of 1-accuracy patterns, the peak of 0-accuracy is higher for DT-DC than for discriminant chronicles. This means that the proportion of patterns that always make wrong decisions is higher

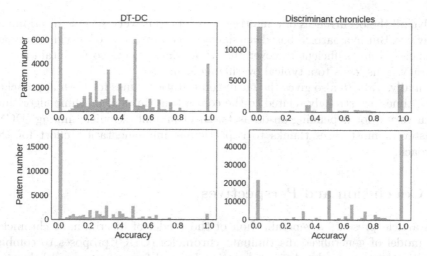

Fig. 5. Accuracy distribution for DT-DC and discriminant chronicles with parameters $\sigma_{min} = 0.5$, $g_{min} = 2$ and $max\_size = 5$ for the first row and $\sigma_{min} = 0.3$, $g_{min} = 2$ and $max\_size = 3$ for the second one.

Table 5. Best accuracy results in SVM-DC-based and discriminant chronicle-based classification and computation times with $\sigma_{min} = 0.4$, $max\_size = 3$ and $g_{min} = 2$.

Dataset	SVM-DC		Discriminant chronicles	
	Accuracy	CPU time (s)	Accuracy	CPU time (s)
asl-bu	**0.73(±0.05)**	**15.2(±0.34)**	0.68(±0.06)	16.4(±0.37)
asl-gt	**0.42(±0.02)**	386(±8.37)	0.32(±0.01)	**7.70(±0.22)**
blocks	0.98(±0.05)	25.4(±2.58)	**1.00(±0.00)**	**11.0(±0.13)**

for DT-DC than for discriminant chronicles and, thus, that DT-DC overfit more the datasets than discriminant chronicles. The same behavior is observed in most of the experiments.

Finally, Table 5 shows the accuracy of SVM-DC and discriminant chronicles for real datasets: *asl-bu*, *asl-gt* and *blocks*. The parameters used for these results were obtained through a grid search. The involved parameters are σ_{min}, g_{min} and γ but also the C parameter of the global linear SVM classifier. Table 5 shows that SVM-DC produces patterns with better accuracy than discriminant chronicles on *asl-bu* and *asl-gt*. For *blocks*, discriminant chronicles are not more accurate than SVM-DC but the discriminant chronicles are discriminant enough to describe such a simple dataset. A classifier based on chronicles is perfect to classify the *blocks* sequences.

These results show that combining decisions of discriminant chronicles makes discriminant less competitive than SVM-DC, even if discriminant chronicles are individually accurate. Thereby, we cannot conclude from previous results that discriminant chronicles are the most accurate GDC. Indeed, chronicles do not involve all the occurrences of a multiset and represent very specific discriminant behaviors. But in a pattern-based classification context, a set of very discriminant chronicles is not sufficient to cover the whole dataset and so to obtain a good accuracy. This leads to a typical overfitting situation.

Finally, Table 5 also gives the mean computation times for both approaches. These times are strongly related to the computing times of machine algorithms which vary a lot depending on datasets. Discriminant chronicle mining (DCM) is faster in most cases thanks to a particular implementation effort for this approach.

7   Conclusion and Perspectives

This article presents a generalization of the model of discriminant chronicles. The model of generalized discriminant chronicles (GDC) proposes to combine a multiset pattern and a decision function learned from the temporal duration between occurrences of a multiset pattern. Initially, discriminant chronicles were extracted using a rule learner and their temporal boundaries were intervals. Such a representation may be too restrictive an assumption on how to discriminate temporal sequences and, thus, had to be generalized.

We demonstrate the expressiveness of the framework by showing that it can model classical patterns (episodes, sequential patterns and chronicles) and episodes, sequential patterns and new types of patterns. DT-DC are based on decision tree classifiers and SVM-DC are based on a SVM classifier.

The experiments show that individual chronicles have good accuracy but SVM-DC overtakes the combination of chronicles on real datasets. An interesting perspective of this work is to blend different types of chronicles within the same combination. Furthermore, comparison in terms of interpretability between several GDC instances would be interesting. Indeed, chronicles are attractive for its interpretability, thanks to its graphical representation. However, new temporal patterns like DT-DC or SVM-DC can not be graphically represented so simply. Then it would be possible to suggest GDC instances that would offer a tradeoff between prediction accuracy and interpretability.

Acknowledgements. This work has received funding from the European Research Council (ERC) under the European Union's Horizon 2020 research and innovation program (grant agreement No [694980] SYNTH: Synthesising Inductive Data Models).

References

1. Asker, L., Boström, H., Karlsson, I., Papapetrou, P., Zhao, J.: Mining candidates for adverse drug interactions in electronic patient records. In: Proceedings of the International Conference on PErvasive Technologies Related to Assistive Environments (PETRA), pp. 22:1–22:4 (2014)
2. Atzmueller, M.: Subgroup discovery. Wiley Interdisc. Rev. Data Min. Knowl. Disc. 5(1), 35–49 (2015)

3. Bringmann, B., Nijssen, S., Zimmermann, A.: Pattern-based classification: a unifying perspective. In: Proceedings of the LeGo Workshop "From Local Patterns to Global Models", p. 10 (2009)
4. Cohen, W.W.: Fast effective rule induction. In: Proceedings of the International Conference on Machine Learning, pp. 115–123 (1995)
5. Cram, D., Mathern, B., Mille, A.: A complete chronicle discovery approach: application to activity analysis. Expert Syst. **29**(4), 321–346 (2012)
6. Dauxais, Y., Guyet, T., Gross-Amblard, D., Happe, A.: Discriminant chronicles mining - application to care pathways analytics. In: Proceedings of 16th Conference on Artificial Intelligence in Medicine (AIME), pp. 234–244 (2017)
7. Dong, G., Li, J.: Efficient mining of emerging patterns: discovering trends and differences. In: Proceedings of the International Conference on Knowledge Discovery and Data Mining (KDD), pp. 43–52 (1999)
8. Fabrègue, M., et al.: Discriminant temporal patterns for linking physico-chemistry and biology in hydro-ecosystem assessment. Ecol. Inf. **24**, 210–221 (2014)
9. Fradkin, D., Mörchen, F.: Mining sequential patterns for classification. Knowl. Inf. Syst. **45**(3), 731–749 (2015). https://doi.org/10.1007/s10115-014-0817-0
10. Giannotti, F., Nanni, M., Pedreschi, D.: Efficient mining of temporally annotated sequences. In: Proceedings of the International Conference on Data Mining (ICDM), pp. 348–359 (2006)
11. Höppner, F.: Discovery of temporal patterns. In: De Raedt, L., Siebes, A. (eds.) PKDD 2001. LNCS (LNAI), vol. 2168, pp. 192–203. Springer, Heidelberg (2001). https://doi.org/10.1007/3-540-44794-6_16
12. Lipton, Z.C., Berkowitz, J., Elkan, C.: A critical review of recurrent neural networks for sequence learning. arXiv preprint arXiv:1506.00019 (2015)
13. Mabroukeh, N.R., Ezeife, C.I.: A taxonomy of sequential pattern mining algorithms. ACM Comput. Surv. (CSUR) **43**(1), 3:1–3:41 (2010)
14. Mannila, H., Toivonen, H., Verkamo, A.I.: Discovery of frequent episodes in event sequences. Data Min. Knowl. Disc. **1**(3), 259–289 (1997). https://doi.org/10.1023/A:1009748302351
15. Nanni, M., Rigotti, C.: Extracting trees of quantitative serial episodes. In: Džeroski, S., Struyf, J. (eds.) KDID 2006. LNCS, vol. 4747, pp. 170–188. Springer, Heidelberg (2007). https://doi.org/10.1007/978-3-540-75549-4_11
16. Pedregosa, F., et al.: Scikit-learn: machine learning in Python. J. Mach. Learn. Res. **12**, 2825–2830 (2011)
17. Quinlan, J.R.: Learning decision tree classifiers. ACM Comput. Surv. (CSUR) **28**(1), 71–72 (1996)
18. Srikant, R., Agrawal, R.: Mining sequential patterns: generalizations and performance improvements. In: Apers, P., Bouzeghoub, M., Gardarin, G. (eds.) EDBT 1996. LNCS, vol. 1057, pp. 1–17. Springer, Heidelberg (1996). https://doi.org/10.1007/BFb0014140
19. Yen, S.-J., Lee, Y.-S.: Mining non-redundant time-gap sequential patterns. Appl. Intell. **39**(4), 727–738 (2013). https://doi.org/10.1007/s10489-013-0426-8

Demand Forecasting in the Presence of Privileged Information

Mozhdeh Ariannezhad[1]([✉]), Sebastian Schelter[2,3], and Maarten de Rijke[2,3]

[1] AIRLab, University of Amsterdam, Amsterdam, The Netherlands
m.ariannezhad@uva.nl
[2] University of Amsterdam, Amsterdam, The Netherlands
{s.schelter,m.derijke}@uva.nl
[3] Ahold Delhaize, Zaandam, The Netherlands

Abstract. Predicting the amount of sales in the future is a fundamental problem in the replenishment process of retail companies. Models for forecasting the demand of an item typically rely on influential features and historical sales of the item. However, the values of some influential features (to which we refer as *non-plannable features*) are only known during model training (for the past), and not for the future at prediction time. Examples of such features include sales in other channels, such as other stores in chain supermarkets. Existing forecasting methods ignore such non-plannable features or wrongly assume that they are also known at prediction time. We identify non-plannable features as privileged information, i.e., information that is available at training time but not at prediction time, and design a neural network to leverage this source of data accordingly. We present a dual branch neural network architecture that incorporates non-plannable features at training time, with a first branch to embed the historical information, and a second branch, the *privileged information* (PI) *branch*, to predict demand based on privileged information. Next, we leverage a single branch network at prediction time, which applies a simulation component to mimic the behavior of the PI branch, whose inputs are not available at prediction time. We evaluate our approach on two real-world forecasting datasets, and find that it outperforms state-of-the-art competitors in terms of mean absolute error and symmetric mean absolute percentage error metrics. We further provide visualizations and conduct experiments to validate the contribution of different components in our proposed architecture.

1 Introduction

Demand forecasting aims to predict future sales and has the potential to significantly improve supply chain management. An accurate forecast prevents overstocking and reduces costs, waste, and losses. At the same time, it also avoids understocking and thereby helps to prevent unfulfilled orders and unsatisfied customers [7,29]. In practice, demand forecasting is modeled as a time series forecasting (TSF) problem, where the goal is to predict future sales volumes

© Springer Nature Switzerland AG 2020
V. Lemaire et al. (Eds.): AALTD 2020, LNAI 12588, pp. 46–62, 2020.
https://doi.org/10.1007/978-3-030-65742-0_4

based on historical sales and influential features [5]. Following the success of deep neural networks in sequence-to-sequence tasks such as machine translation [28], recent work has studied the effectiveness of neural networks for TSF in general [16,21,23], and demand forecasting in particular [8,11,20].

We divide influential features into two categories, based on their availability at the time of forecast. *Plannable features* are known for the past and the future; examples are time-dependent features such as calendar events, or static features such as product characteristics. *Non-plannable features* are time-dependent and only known for the past, e.g., sales in other stores. Many neural-based approaches use an encoder-decoder structure to map the history of a time series to its future. A common strategy when treating influential features is to feed their historical values to the encoder, either alongside the historical sales data [11], or to a different layer that is responsible for encoding them [8]. The decoder then either uses their future values to produce the forecast [11,31], or assumes that they are unknown [8]. Neither of these schemes is ideal for non-plannable features. Using future values of influential features in the decoding stage makes sense for plannable features, but the approach is not applicable for non-plannable features. A model trained in that way is not applicable in a real world setting as the non-plannable features are not known at prediction time. Simply ignoring non-plannable features – both at training and prediction time – is not optimal either as they carry important information that should be leveraged at the time of training the model.

In this paper, we therefore propose an indirect approach to model the effects of non-plannable features, and treat them as *privileged information* [18], i.e., information that is available at the time of training the model but not at prediction time (Sect. 3.1). We introduce a neural network architecture to capture the effect of these features at training time, and use this effect to produce a forecast at prediction time (Sect. 3.2). To this end, we propose *two different network architectures for training and prediction*. At the time of training, the network has two different branches. The first branch is responsible for modeling the effect of historical sales and plannable features, while the second branch uses non-plannable features as input to produce a forecast based on them. At prediction time, the second branch is not available, and is *replaced by a simulation network*, which is trained to mimic its behavior (Sect. 3.3).

Our experimental evaluation demonstrate that the proposed model outperforms state-of-the-art baselines on several datasets in terms of mean absolute error and symmetric mean absolute percentage error. Our experiments show that the proposed network is not only able to learn from non-plannable features at training time, but can also embed this type of information for use at prediction time.

We summarize the contributions of our research as follows:

– We categorize the influential features for demand forecasting into two categories, based on their availability for the time of forecast. We propose to treat the non-plannable features as privileged information (Sect. 3.1).

– We design a novel neural network architecture that is able to leverage privileged information. To this end, we propose to use two different networks for training and prediction time, where the network at prediction time simulates the effect of the unknown privileged information (Sect. 3.2).
– We conduct extensive experiments on two publicly available datasets. Our experimental results show that our approach outperforms state-of-the-art baselines for demand forecasting in the majority of cases in terms of mean absolute error and symmetric mean absolute percentage error metrics (Sect. 5).

2   Related Work

Time Series Forecasting. Prior work on time series forecasting (TSF) mostly focuses on linear approaches [22], such as Auto-Regressive Integrated Moving Average (ARIMA) model [6], with a solid underlying theory and relatively few parameters. However, linear methods cannot model non-linear temporal dependencies and complex relationships between different dimensions of a time series. Recently, neural network-based approaches have found their way into TSF. The most dominant type of network used are recurrent neural networks (RNNs) and long short-term memory networks (LSTMs) [2,4,21,23,25]. Convolutional neural networks (CNNs) are also considered in the literature [14,24], and some work studies networks with both recurrent and convolutional components [16,32].

For the task of demand forecasting, state-of-the-art approaches mostly employ neural-based models. TADA [8] uses different LSTM layers to model different kinds of influential features, and the multimodal-attention model proposed in [11] uses a bidirectional LSTM with an attention mechanism to better capture latent patterns in historical data. To incorporate the impact of substitutable products with respect to the target product, DSF [20] uses a sequence-to-sequence structure with gated recurrent units.

Our focus is on modeling non-plannable features. In previous work, non-plannable features are either treated as plannable, i.e., with the unrealistic assumption that their future values are known at prediction time, or are only used as historical data. None of these approaches is able to incorporate non-plannable features in a realistic manner. In contrast, we propose a neural architecture, that (1) is capable of modeling non-plannable features, and (2) has different components for historical sales and influential features, which are usually treated in the same way in previous work.

Learning Under Privileged Information. The learning under privileged information (LUPI) framework was originally proposed for support vector machines [30]. The idea was again popularized in [18], where it was unified with knowledge distillation for neural networks. Most of the work done in this area is in the field of computer vision [9,12,17], with teacher-student networks as the dominant approach. Teacher-student networks are mostly based on distilling knowledge from a 'teacher' network to a 'student' network at training time through the loss function, which makes sense for classification problems [18], where class probabilities produced by the teacher network are treated as 'soft targets' for training the student network.

To the best of our knowledge, LUPI has never been used in time series forecasting and utilizing existing LUPI frameworks is not straight-forward in a forecasting scenario. In this work, we propose a network architecture to leverage non-plannable features. We achieve this with two different networks at training and prediction time. In contrast to common teacher-student networks, our second network is not guided by a loss component, but learns a simulation component that mimics the output of the missing branch.

3   A Privileged Information-Aware Neural Network

We present a dual branch neural network architecture for demand forecasting. It incorporates non-plannable features as privileged information (PI) at training time, with a first branch (the *historical branch*) that embeds historical information and plannable features via dilated causal convolutional layers, and a second branch, the *PIBranch*, which leverages fully-connected feed-forward layers to predict sales based on privileged information. At prediction time, we apply a slightly different single branch network with a simulation component to mimic the behavior of the PIBranch, whose inputs are not available at prediction time.

3.1   Problem Statement

The goal of a demand forecasting model is to predict the amount of sales in the future. Many different settings exist for building a forecasting model. Without loss of generality, we consider the case of multiple stores and multiple items, and forecast the demand of each item per store. Formally, for an arbitrary target product in a target store, the goal of the forecasting model is to predict

$$\{\hat{y}_t\}_{t=T+1}^{T+\Delta} = \{\hat{y}_{T+1}, \ldots, \hat{y}_{T+\Delta}\},$$

where T is the length of history being considered, Δ is the forecast horizon, and $\hat{y}_t \in \mathbb{R}$ denotes the predicted sales of the target item in the target store at time t. Future sales are affected by both the history of sales in the past, and other influential features. Influential features can be divided into two categories. *Plannable features* are known both for the past and the future; they can be static, such as product-dependent features like category and brand, or time-dependent, such as promotional campaigns and calendar events. Future values of *non-plannable features* are not known at the time of forecast; examples include behavior data from users on an online shopping website, sales of similar items in the same store, or sales of the same item in other stores.

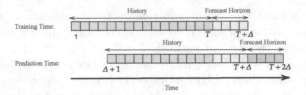

Fig. 1. The 'walk-forward' training scheme. A history of length T is used for training, and validation is performed on $[T+1, T+\Delta]$. For predication time, the history is shifted Δ steps, and $[T + \Delta + 1, T + 2\Delta]$ is used as forecast horizon.

We design a forecasting model that incorporates both types of feature alongside the historical sales. Formally:

$$\{\hat{y}_t\}_{t=T+1}^{T+\Delta} = F\left(\{y_t\}_{t=1}^{T}, \{\mathbf{x}_t^p\}_{t=1}^{T+\Delta}, \{\mathbf{x}_t^{np}\}_{t=1}^{T}\right), \tag{1}$$

where $F(\cdot)$ is a non-linear mapping function that we learn, and

$$\{y_t\}_{t=1}^{T} = \{y_1, y_2, \ldots, y_T\} \tag{2}$$
$$\{\mathbf{x}_t^p\}_{t=1}^{T+\Delta} = \{\mathbf{x}_1^p, \mathbf{x}_2^p, \ldots, \mathbf{x}_{T+\Delta}^p\} \tag{3}$$
$$\{\mathbf{x}_t^{np}\}_{t=1}^{T} = \{\mathbf{x}_1^{np}, \mathbf{x}_2^{np}, \ldots, \mathbf{x}_T^{np}\} \tag{4}$$

denote the historical sales of the target item, and the corresponding plannable features $\mathbf{x}^p \in \mathbb{R}^n$ and non-plannable features $\mathbf{x}^{np} \in \mathbb{R}^m$ for the item, respectively, with the corresponding feature dimensions n and m.

For training our model, we adopt the 'walk-forward' training schema that is a common choice for time series data [8,15,27]. In this approach, which we illustrate in Fig. 1, we apply a sliding window to divide the data into history and forecast horizon, and shift this sliding window Δ steps from training time to prediction time. In other words, assuming the length of the whole dataset is $T + 2\Delta$ and $T \gg \Delta$ is the length of the sliding window, the $[1, T]$ interval is used as the history training time, the $[T + 1, T + \Delta]$ interval is used as the forecast horizon at training time (i.e., the validation window), the $[\Delta + 1, T + \Delta]$ interval is used as the history at prediction time and finally, the $[T + \Delta + 1, T + 2\Delta]$ interval is used as the forecast horizon at prediction time.

We evaluate our model on the forecast horizon at prediction time, for which the non-plannable features are not available. For training and hyperparameter selection, we leverage the history and the validation window, for which non-plannable features are known. This assumption is inline with real world cases, where a forecasting model is trained on the past data, for which the values of all influential features are already known. We rephrase Eq. 1 into Eq. 5 at training time and into Eq. 6 at prediction time in order to account for this setup:

$$\{\hat{y}_t\}_{t=T+1}^{T+\Delta} = F_1\left(\{y_t\}_{t=1}^{T}, \{\mathbf{x}_t^p\}_{t=1}^{T+\Delta}, \{\mathbf{x}_t^{np}\}_{t=1}^{T+\Delta}\right) \tag{5}$$
$$\{\hat{y}_t\}_{t=T+\Delta+1}^{T+2\Delta} = F_2\left(\{y_t\}_{t=\Delta+1}^{T+\Delta}, \{\mathbf{x}_t^p\}_{t=\Delta+1}^{T+2\Delta}\right), \tag{6}$$

where $F_1(\cdot)$ is the function that we learn at training time and $F_2(\cdot)$ is the function that we apply at prediction time to produce the forecast.

3.2 Architecture Overview

We model non-plannable features as privileged information (PI), i.e., information that is available at the training time but not available at prediction time. This approach requires different forecasting models for training and prediction time. We therefore propose: a *Dual Branch PI Aware Neural Network* (DB-PIANN) to embed both the historical sales and PI at training time, and a *Single Branch PI Aware Neural Network* (SB-PIANN) to produce the forecast at prediction time. Figure 2a illustrates the architecture of DB-PIANN. Historical sales and plannable features are fed into one branch of the network, while privileged information is fed into a different branch. The outputs of these two branches are then merged with a combination layer, and fed into the output module to produce the forecast. The unavailablity of the PI for the future implies that we cannot produce a forecast with DB-PIANN at prediction time. However, only the input to the privileged information branch (PIBranch) is missing at prediction time, and we can still utilize the historical branch (HiBranch) of DB-PIANN.

We tackle this challenge by *training an additional simulation network to mimic the behavior of the missing PI Branch*. This network takes the output of the historical branch as input, and learns to reproduce the output of the combination layer. Figure 2b details the architecture of the SB-PIANN. The difference to DB-PIANN is that the PIBranch and the combination layer are replaced with the simulation network; the other branch is identical. DB-PIANN and SB-PIANN model $F_1(\cdot)$ and $F_2(\cdot)$ in Eq. 5 and Eq. 6, respectively.

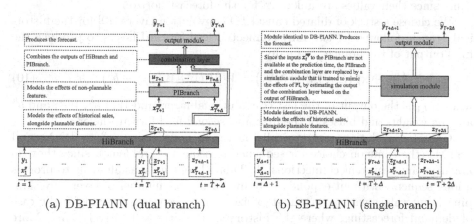

(a) DB-PIANN (dual branch)   (b) SB-PIANN (single branch)

Fig. 2. Neural network architectures for PI-aware demand forecasting. Figure 2a illustrates the architecture of DB-PIANN applied at training time and Fig. 2b details the architecture of the SB-PIANN, which we leverage at prediction time

3.3   Architecture Details

Next, we introduce the details of DB-PIANN and SB-PIANN. As illustrated in Fig. 2a, DB-PIANN consists of four modules: (1) a *historical branch* (HiBranch), which is responsible for modeling the effects of historical sales data, along with plannable features, (2) a *privileged information branch* (PIBranch), which takes care of non-plannable features, (3) a *combination layer*, which combines the outputs of the two previous branches, and (4) an *output module*, which produces the final forecasts.

Based on these definitions, we break up Eq. 5 as follows:

$$\{\hat{y}_t\}_{t=T+1}^{T+\Delta} = F_1(\{c_t\}_{t=T+1}^{T+\Delta}), \tag{7}$$

where F_1 is the output module, and $\{c_t\}_{t=T+1}^{T+\Delta}$ is the output of the combination layer, defined as:

$$\{c_t\}_{t=T+1}^{T+\Delta} = \{z_t + u_t\}_{t=T+1}^{T+\Delta}, \tag{8}$$

where $\{z_t\}_{t=T+1}^{T+\Delta}$ and $\{u_t\}_{t=T+1}^{T+\Delta}$ are the outputs of the HiBranch and PIBranch, respectively (Fig. 2a).

Historical Branch. The input of this branch, namely N_1, consists of the historical sales of an item y_t with of length T, and the plannable features $\mathbf{x_t^P}$ of length $T + \Delta$ corresponding to that item. The output of this branch, z_t, has the length of Δ, and will be fed to the combination layer. Formally:

$$\{z_t\}_{t=T+1}^{T+\Delta} = N_1(\{y_t \oplus \mathbf{x_t^P}\}_{t=1}^{T}, \{\mathbf{x_t^P}\}_{t=T+1}^{T+\Delta}), \tag{9}$$

where $\oplus$ denotes the concatenation operation. With this definition, HiBranch can be applied in SB-PIANN at prediction time, where we shift the time Δ steps. However, one cannot include non-plannable features as well at prediction time, since their values are unknown for the forecast horizon.

We choose a stack of dilated causal 1D convolution layers [19] for the historical branch, followed by two fully-connected feed-forward layers to produce the final output of the branch. Formally:

$$N_1(\{y_t \oplus \mathbf{x_t^P}\}_{t=1}^{T}, \{\mathbf{x_t^P}\}_{t=T+1}^{T+\Delta}) = W_2^h \text{ReLU}(W_1^h c_{L_1} + b_1^h) + b_2^h, \tag{10}$$

where c_{L1} is the output of the last convolutional layer, and W_1^h, b_1^h, W_2^h, and b_2^h are the weights and biases of the first and second feed-forward layer.

Causal convolutions are typically faster to train compared to RNNs and LSTMs (the common choice for sequence to sequence problems), since they do not have any recurrent connections [3]. Dilated convolutions allow us to process a long sequence without exponentially increasing the number of layers, by skipping input values with a certain step size. This is especially useful for the case of demand forecasting, where the history of the data is relatively long. More formally, for a sequence input x and a filter f with size k, a causal dilated convolution operation g on element s of the sequence is defined as:

$$g(s) = (x *_d f)(s) = \sum_{i=0}^{k-1} f(i) \cdot x_{s-d \cdot i}, \tag{11}$$

where d is the dilation factor. Following [19], we increase d exponentially with the depth of the network, i.e., $d = O(2^j)$ at level j of the network, where $j \in \{0, \ldots, L_1 - 1\}$ and L_1 denotes the number of layers. All parameters are shared across the whole forecast horizon.

The historical branch N_1 operates on a rolling basis, meaning that the output $z_t \in \mathbb{R}$ is produced one step at a time and is concatenated to the history in the input, which is is then shifted one step forward, and fed back into the first convolutional layer, until the end of the forecast horizon is reached (see Fig. 2a).

Privileged Information Branch. The input of this branch, namely N_2, are the non-plannable features with a length of Δ, and the output of this branch, u_t, (also of length Δ), will be fed to the combination layer. To make use of all the available training data, we apply a sliding window of length Δ and move it forward one step at a time, until we reach the end of the training interval, i.e., $T + \Delta$. For the PIBranch N_2, we leverage fully-connected feed-forward layers with dropout [26] applied after each hidden layer. We use a ReLU activation function for the hidden layers, and a linear fully-connected layer for the output. Formally:

$$\{u_t\}_{t=i}^{i+\Delta} = N_2(\{\mathbf{x^{nP}}_t\}_{t=i}^{i+\Delta}) = W_{L_2}^{np}\text{ReLU}(W_l^{np}x_l + b_l^{np}) + b_{L_2}^{np}, \quad (12)$$

where $1 \leqslant i \leqslant T$ is the start of the sliding window, L_2 denotes the number of layers, and W_l^{np} and b_l^{np} are the weights and biases of the l-th layer for $l \in \{1, \ldots, L_2\}$; x_l is the input of the l-th layer, which is the output of the previous layer for $l \in \{2, \ldots, L_2\}$, and equal to $\{\mathbf{x^{nP}}_t\}_{t=T+1}^{T+\Delta}$ for $l = 1$.

Note that the PIBranch cannot be used in SB-PIANN at prediction time, since its input values are unknown for the forecast horizon.

Output Module. In SB-PIANN, this module consumes the output of the combination layer, and produces the predicted sale values for the validation period. We again apply fully-connected feed-forward layers:

$$\{\hat{y}_t\}_{t=T+1}^{T+\Delta} = F_1(\{u_t + z_t\}_{t=T+1}^{T+\Delta}) = W_{L_3}^o\text{ReLU}(W_l^o s_l + b_l^o) + b_{L_3}^o, \quad (13)$$

where W_l^o and b_l^o are the weights and biases of the l-th layer for $l \in \{1, \ldots, L_3\}$, s_l is the input of l-th layer, which is the output of the previous layer for $l \in \{2, \ldots, L_3\}$, and equal to $\{u_t + z_t\}_{t=T+1}^{T+\Delta}$ for $l = 1$.

Together, the definitions of the aforementioned modules complete the architecture of DB-PIANN, according to Eq. 7.

SB-PIANN. Next, we define our second architecture SB-PIANN, which is going to produce the forecast at prediction time. We break down Eq. 6 as follows:

$$\begin{aligned}
\{\hat{y}_t\}_{t=T+\Delta+1}^{T+2\Delta} &= F_2\left(\{y_t\}_{t=\Delta+1}^{T+\Delta}, \{\mathbf{x^P}_t\}_{t=\Delta+1}^{T+2\Delta}\right) \\
&= F_1\left(S(\{z_t\}_{t=T+\Delta+1}^{T+2\Delta})\right) \quad (14) \\
&= F_1\left(S(N_1(\{y_t \oplus \mathbf{x_t^P}\}_{t=\Delta+1}^{T+\Delta}, \{\mathbf{x_t^P}\}_{t=T+\Delta+1}^{T+2\Delta}))\right),
\end{aligned}$$

where S is the simulation module, which we define in Eq. 15 below, N_1 refers to the historical branch defined in Eq. 10, and F_1 depicts the output module, defined in Eq. 13.

Simulation Module. The purpose of the *simulation module* is to replace the missing PIBranch of DB-PIANN. As we cannot use the PIBranch at prediction time, one of the inputs of the combination layer is not available at prediction timeas well (see Fig. 2). Therefore, we train a feed-forward neural network to estimate the output of the combination layer based on the output of HiBranch. Formally:

$$S(\{z_t\}_{t=T+\Delta+1}^{T+2\Delta}) = W_{L_4}^s \text{ReLU}(W_l^s p_l + b_l^s) + b_{L_4}^s, \tag{15}$$

where W_l^s and b_l^s are the weights and biases of the l-th layer for $l \in \{1, \ldots, L_4\}$. p_l is the input of l-th layer, which is the output of the previous layer for $l \in \{2, \ldots, L_4\}$, and equal to $\{z_t\}_{t=T+\Delta+1}^{T+2\Delta}$ for $l = 1$.

3.4 Learning Process

Many options exist for training a forecasting model for a collection of time series, such as the sales per item per store in our case. Following previous work [8,23], we train a single model for all of the items; each training sample contains the sales of a single item in a single store, along with its corresponding influential features.

Table 1. Dataset statistics.

Dataset	Items	Stores	#time series	Time series' length
Favorita	1,656	54	11,614	365
Dunnhumby	1,101	26	8,825	117

We train the PIBranch N_2 and the historical branch N_1 separately, and then train DB-PIANN using the learned models, as outlined in the previous section. We subsequently train the simulation network S in isolation. With this scheme, SB-PIANN does not actually need training and can be composed from the previously learned modules, i.e., HiBranch, the simulation module and the output module. We leverage the mean absolute error between the predicted values and the actual values as objective function to train all the components of our architecture. All of our proposed modules are smooth and differentiable, which allows us to learn their parameters by standard back propagation. With mean absolute error as the objective function, we define the loss for each module as follows:

$$\mathcal{L} = \frac{1}{N}\Big(\sum_{n=1}^{N}\sum_{t=T+1}^{T+\Delta} |y_{nt}^a - y_{nt}^p|\Big), \tag{16}$$

where N is the number of training samples, n is an index for the samples, y_{nt}^a is the label for sample n at time t, and y_{nt}^p is the output of the corresponding module. Specifically, when training the HiBranch, PIBranch, and output module, y_{nt}^a is equal to the actual sales of the training sample n at time t, and y_{nt}^p is equal to the corresponding value of z_t, u_t, and $\hat{y}_t$ for sample n. For the simulation module, y_{nt}^a translates to the corresponding values of $z_t + u_t$ of sample n.

4   Experimental Setup

Datasets. Unfortunately, most of the sales datasets from existing work are not publicly available. For example, we could neither obtain the 'One Stop Warehouse' dataset from [8], the Amazon Demand Forecasting dataset from [31], nor the JD50K Online Sales Forecasting dataset from [11].

We therefore evaluate the performance of SB-PIANN on two publicly available datasets to ensure reproducibility. Both datasets contain sales numbers at the product and store level; we therefore set the goal of predicting the sales per product per store for a certain forecast horizon. Here, for a target item in a target store, we treat its sales in other stores as the privileged information. The *Favorita* dataset contains daily sales of thousands of items across 54 stores located in Ecuador.[1] We use the data from the 15th of August 2016 to the 15th of August 2017, and only consider items that have less than five days of sales data missing.[2] We also use the *The Complete Journey* dataset published by *Dunnhumby*.[3] This dataset contains around 300 million transactions for ~5,000 items across ~760 distinct stores, spanning more than two years of history. We randomly select a subset of items and stores for our experiments, and aggregate sales on a weekly basis to reduce the sparsity. The statistics of the datasets are shown in Table 1. Not all of the items are sold in all of the stores, so the total number of time series, i.e., training samples, is different from the number of items multiplied by the number of stores.

Influential Features. Our design of the network architecture does not restrict the types of influential features that our approach can incorporate. However, we work with non-plannable features only in the experiments, as their effect on the forecast is the focus of this paper. For such non-plannable features, we rely on the sales of an item in other stores. In many demand forecasting datasets, including both of the datasets that we use, the retail company has more than one store, and the sales of a target item in other stores can help to forecast the demand in the target store. Formally:

$$\mathbf{x^{nP}}_t = \{(x_t^{s_0}, x_t^{s_1}, \ldots, x_t^{s_n})\}, \tag{17}$$

where n is the number of stores and $x_t^{s_i}$ denotes the sales for the target item in store s_i at time t.

---

[1] https://www.kaggle.com/c/favorita-grocery-sales-forecasting/data.

[2] The same dataset is used in [8], however the authors do not mention how they created a subset of the original dataset in their paper. Nevertheless, we achieve a comparable performance with their method on our version of the data.

[3] https://www.dunnhumby.com/careers/engineering/sourcefiles.

Training and Testing. Analogous to previous work, we use the walk-forward strategy [8,27] for the train-test split, as detailed in Sect. 3.3. Following this strategy, we split the data according to the time dimension, and not based on stores and items. Given a time series with a total size of T for the history of sales data and Δ steps to predict, we use $[1, T]$ as the training data, $[T + 1, T + \Delta]$ for validation and $[T + \Delta + 1, T + 2\Delta]$ for testing (as illustrated in Fig. 1). We experiment with $\Delta \in \{2, 8, 16\}$.

Implementation Details. Our models are implemented using the Keras framework [10] with TensorFlow as backend [1]. We use mini-batch stochastic gradient descent (SGD) together with the Adam optimizer [13] to train the models. We set the batch size to 32, and stop training when the loss on the validation set converges. All the methods run on a Linux server with an Intel Xeon CPU, and a GeForce GTX 980 (Maxwell GM204) GPU. The GPU code is implemented using CUDA 9.

Metrics. We consider two metrics for evaluation: mean absolute error (MAE) $= \frac{1}{N \times \Delta} \sum_{n=1}^{N} \sum_{t=T+\Delta+1}^{T+2\Delta} |y_t - \hat{y}_t|$ and symmetric mean absolute percentage error (SMAPE) $= \frac{100}{N \times \Delta} \sum_{n=1}^{N} \sum_{t=T+\Delta+1}^{T+2\Delta} \frac{|y_t - \hat{y}_t|}{(|y_t| + |\hat{y}_t|)/2}$, where y_t and $\hat{y}_t$ denote real and predicted sales, respectively.

Parameter Settings. For all of the parameters of the networks, we conduct a grid search and leverage the parameters with the best performance on the validation set. For the feed-forward modules, we conduct a grid search over $\{1, 2, 3, 4\}$ for the number of hidden layers, $\{16, 32, 64, 128, 256, 512\}$ for the number of nodes in each hidden layer, and $\{0.1, 0.2, 0.3, 0.4\}$ for the dropout rate. For the convolutional layers, we search over $\{16, 32, 64\}$ for the number of filters, $\{2, 4, 8, 16\}$ for the filter size, and $\{4, 5, 6, 7, 8\}$ for the number of layers.

5   Experimental Results

5.1   Capturing the Effects of Privileged Information

Our first set of experiments addresses the following research question: *To what extent is the proposed model, SB-PIANN, able to capture the effect of privileged information?* We conduct experiments with different components of our proposed model to answer our first research question. The purpose of these experiments is to reveal the importance of privileged information and to quantify its impact on the final performance of SB-PIANN. We compare its performance to that of the HiBranch and PIBranch in isolation, as well to the "oracle" performance of DB-PIANN. DB-PIANN's performance is only reported for the sake of comparison; the model cannot be used for prediction in real world cases due to the unavailability of the values of non-plannable features at prediction time.

Results and Discussion. Table 2 shows the performance of HiBranch, PIBranch, SB-PIANN and DB-PIANN in terms of MAE and SMAPE, on testing time intervals, i.e., $[T + \Delta + 1, T + 2\Delta]$. The results are reported for

Table 2. Performance of SB-PIANN and DB-PIANN and their components on the Favorita and Dunnhumby datasets. * indicates that a component is significantly outperformed by SB-PIANN; ⁻ indicaties that there is no significant difference between DB-PIANN and SB-PIANN (student t-test, $\alpha = 0.05$).

Method	$\Delta = 2$		$\Delta = 8$		$\Delta = 16$	
	MAE	SMAPE	MAE	SMAPE	MAE	SMAPE
Favorita dataset						
HiBranch	6.126*	38.840*	6.494*	38.563*	7.278*	38.794*
PIBranch	6.415*	38.235*	6.664*	39.692*	7.080*	38.431*
SB-PIANN	6.069	37.139	6.277	38.102	6.868	38.241
DB-PIANN	5.985⁻	36.956⁻	6.052⁻	37.304⁻	6.752⁻	37.314⁻
Dunnhumby dataset						
HiBranch	3.593	43.042*	3.661*	44.348*	4.079*	50.289*
PIBranch	3.669*	44.108*	3.640	44.006	3.691	45.059
SB-PIANN	3.558	42.567	3.612	43.506	4.059	49.540
DB-PIANN	3.555⁻	42.531⁻	3.601⁻	43.384⁻	3.899⁻	47.817⁻

all Δ settings on the Favorita and Dunnhumby datasets. We compare the performance of two different branches for different forecast horizons. We observe that the performance of the PIBranch is comparable to that of the HiBranch for shorter forecast horizons, i.e., 2 and 8, and outperforms HiBranch for the longest forecast horizon. This shows that the PIBranch is capable of predicting the sales at least as well as HiBranch, and is more robust with respect to forecast horizon. We attribute the ability of PIBranch to predict future sales to two aspects: First, we note that the historical sales data are not taken into account in the PIBranch, and the forecast is produced solely based on the privileged information. This indicates the usefulness of this information for demand forecasting. Second, the obtained performance points out the effectiveness of the proposed network to model the effects of privileged information. In other words, a deep feed-forward neural network is capable of producing a forecast based on the sales of products in other stores, and this forecast is comparable to and in some cases superior to the one based on the historical sales.

We also report the performance of DB-PIANN, while noting (again) that it cannot be used in a real world setting, as it requires the privileged information, which is not available at prediction time. DB-PIANN outperforms individual branches on both datasets and for all of the forecast horizons, which shows that it is able to leverage both the historical sales and the privileged information to produce the final forecast. The performance gain also indicates that both branches contribute positively to the final forecast; their contribution is not overlapping, and they are both essential to achieve the highest performance.

Finally, we observe that the performance of SB-PIANN is better than that of HiBranch and is relatively close to DB-PIANN, on both datasets and for all forecast horizons. Recall that SB-PIANN is not using the inputs of the PIBranch,

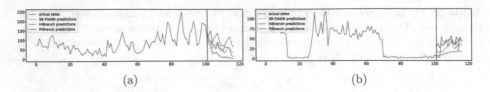

Fig. 3. Forecasts produced by HiBranch, PIBranch and SB-PIANN, along with the actual sales, for sample products. Fig. 3a is taken from Favorita and Fig. 3b is selected from Dunnhumby. y axis shows the sales and x axis is the time. Fig. 3a and Fig. 3b show the improved forecasts using PI.

and indirectly models the effects of privileged information via the simulation component. The close performance of SB-PIANN to that of DB-PIANN shows that the proposed approach is able to embed the PI from train to test time effectively. In other words, the simulation component is capable of transferring the effects of privileged information absorbed in the train time to the prediction time. The superiority of SB-PIANN compared to the HiBranch supports the hypothesis that utilizing the privileged information leads to a better performance than a forecast that only relies on the historical data.

To gain more insights into forecasts by our proposed method, we visualize a few examples in Fig. 3. Each figure shows the demand for a product in a store, taken from both Favorita and Dunnhumby. We show 101 days of history for both datasets. The history of sales as well as the actual sales in the future are plotted, along with forecasts made by PIBranch, HiBranch, and SB-PIANN. We observe that the forecast based on PI has a different trend from the one based on the history of sales. This might be a result of an event that affects the sales of the product in the stores, and leads to a better prediction for the future sales, compared to the forecast of HiBranch, which is based on the history of sales in the target store. We can also recognize that SB-PIANN picks-up on this difference, and its forecast is superior to that of HiBranch.

5.2   Comparison to Existing Approaches for Demand Forecasting

Next, we focus on the following question: *How effective and accurate is SB-PIANN for the task of demand forecasting?* We compare SB-PIANN to the following state-of-the-art neural network-based forecasting models:

DA-RNN [21]. This dual-stage attention-based RNN method is proposed for time series prediction. It uses an encoder-decoder structure with two different attention mechanisms. In this model, the PI are available also for the future. However, this is an unrealistic assumption; we use this model as the baseline to evaluate the ability of our model in making use of PI.

LSTNet [16]. The long- and short-term time-series network uses both CNNs and RNNs to capture both short-term and long-term trending patterns of the time series. It also has an auto-regressive component. The architecture is proposed for multi-variate TSF, therefore we build a model for each item and forecast for the target stores simultaneously. For LSTNet, we only make use of the historical values of PI for both training and test time.

TADA [8]. An encoder-decoder based architecture that uses two LSTM branches for encoding different types of feature. Since the method was tested on the Favorita dataset, we also report their results based on their original division of features for this dataset. For the Dunnhumby dataset, we randomly divide the data from other stores into two categories. We again only make use of the historical values of PI available for both training and test time.

All baselines report superior performance compared to auto-regressive models, decision tree models, and simpler neural network-based approaches. We therefore omit these methods from our evaluation. Aside from their state-of-the-art performance, we chose these methods as baseline to cover a wide range of possible approaches. Specifically, they apply different training schemes; DA-RNN trains a model per time-series, LSTNet trains a model per store, and TADA uses an approach similar to ours, training a single model for all of the data. They also differ in the ways they treat the PI. For DA-RNN, the assumption is that these features are available for both history and the future (which is not true in real world settings), while the other two only use the historical values of the PI.

Implementation Details. For the baseline implementation, we asked the authors of TADA and DA-RNN for the code and they kindly provided us with their own implementation. For LSTNet, we leverage the code which is made publicly available by the authors.[4] We use the same testing environment as SB-PIANN for the baselines, as outlined in Sect. 4.

Table 3. Comparison of neural network-based forecasting methods on the Favorita and Dunnhumby datasets. * indicates that a method is significantly outperformed by SB-PIANN (student t-test, $\alpha = 0.05$).

Method	Uses PI		$\Delta = 2$		$\Delta = 8$		$\Delta = 16$	
	Train	Test	MAE	SMAPE	MAE	SMAPE	MAE	SMAPE
Favorita dataset								
DA-RNN	+	+	9.343*	48.864*	8.583*	46.174*	8.709*	43.132*
LSTNet	−	−	6.619*	40.509*	7.481*	43.920*	8.332*	45.016*
TADA	−	−	6.428*	**36.744**	7.431*	41.389*	8.463*	43.165*
SB-PIANN	+	−	**6.069**	37.139	**6.277**	**38.102**	**6.868**	**38.241**
Dunnhumby dataset								
DA-RNN	+	+	4.535*	52.779*	3.995*	47.154*	4.168*	**48.942**
LSTNet	−	−	4.134*	48.484*	4.818*	55.310*	5.673*	59.864*
TADA	−	−	4.367*	49.018*	5.777*	63.834*	8.251*	78.055*
SB-PIANN	+	−	**3.558**	**42.567**	**3.612**	**43.506**	**4.059**	49.540

---

[4] https://github.com/laiguokun/LSTNet.

Results and Discussion. Table 3 shows the performance of SB-PIANN compared to the baselines for different forecast horizons on Favorita and Dunnhumby, respectively. The best performance is highlighted with bold face. In almost all cases, SB-PIANN outperforms the baselines in terms of MAE and SMAPE. We observe that the performance of all methods starts to drop with the increase of the forecast horizon, but SB-PIANN is more robust with respect to the length of the forecast horizon. In all cases, the performance is better on Favorita than on Dunnhumby in terms of SMAPE. This might be due to the fact that the sales data in Dunnhumby is more scarce.

According to the results, SB-PIANN even outperforms DA-RNN, which makes the unrealistic assumption of having the PI available at prediction time. The fact that SB-PIANN outperforms even DA-RNN shows the effectiveness of our proposed approach to leverage privileged information. The lower performance of DA-RNN might also be a result of its training scheme, i.e., building a forecast model per time series. In this scenario, the model cannot learn from the similarity and differences between different products and stores.

While the special type of PI that we experiment with, i.e., sales in other stores, suggests the use of a multi-variate TSF model, our experiments show that compared to LSTNet, a state-of-the-art model proposed for the multi-variate TSF, SB-PIANN, performs better in all cases. On the Favorita dataset, we compare the performance of SB-PIANN with the original version of TADA, i.e., using the definition that the authors propose for internal and external features. In this version, a set of 13 attributes are used as features, such as the location of the stores and the oil price. In our approach, we only rely on the sales of the items in other stores as features, and we observe that our model performs significantly better in terms of both error metrics, while using no manual categorization of features. The gain in performance is more significant on the Dunnhumby dataset, which indicates the limitation of TADA when the division of features into internal and external as required by TADA is not straightforward. In other words, the performance of TADA degrades when influential features cannot be characterized as internal and external, and accordingly, cannot be fed into the corresponding LSTM layers.

6   Conclusions and Future Work

Demand forecasting is a fundamental problem in the replenishment process of retail companies. Models for forecasting the demand for a product usually consider patterns in the history of sales data and influential features. Such influential features can be divided into two categories: plannable features that are known for the past and the future, and non-plannable features, for which the future values are unknown. Neural forecasting models usually ignore non-plannable features when predicting the amount of sales in the future.

In this paper, we identify non-plannable features as privileged information and design a novel neural network to utilize them. We propose two different network architectures for training and prediction time. At the time of training, the network has two different branches to model the effect of historical sales and non-plannable features. At prediction time, the second branch is not available, and is replaced by a simulation network that is trained to mimic its behavior. We extensively evaluate our proposed approach on two real-world forecasting datasets, and find that it outperforms state-of-the-art baselines in terms of mean absolute error and symmetric mean absolute percentage error metrics.

While our proposed architecture is capable of plannable features, our focus in this paper is on modeling non-plannable features. In future work, it will be appealing to study different approaches to incorporate plannable features in the model; aside from treating them as an extra dimension of the historical sales, they can be fed into the PIBranch, or an extra dedicated branch. Moreover, although we make no specific assumptions about the type of non-plannable features in our model, we rely on a single type of non-plannable features in our experiments. We also aim to investigate the impact of more sources of privileged information.

Code and Data. To facilitate the reproducibility of the reported results, this work only made use of publicly available data and our experimental implementation is publicly available at https://github.com/mzhariann/PIANN.

Acknowledgments. This research was supported by Ahold Delhaize. All content represents the opinion of the authors, which is not necessarily shared or endorsed by their respective employers and/or sponsors.

References

1. Abadi, M., et al.: Tensorflow: a system for large-scale machine learning. In: OSDI, pp. 265–283 (2016)
2. Adhikari, B., Xu, X., Ramakrishnan, N., Prakash, B.A.: EpiDeep: exploiting embeddings for epidemic forecasting. In: KDD, pp. 577 586 (2019)
3. Bai, S., Kolter, J.Z., Koltun, V.: Convolutional sequence modeling revisited. In: ICLR (2018)
4. Bao, W., Yue, J., Rao, Y.: A deep learning framework for financial time series using stacked autoencoders and long-short term memory. PLoS One **12**(7), e0180944 (2017)
5. Boese, J., et al.: Probabilistic demand forecasting at scale. PVLDB **10**(12), 1694–1705 (2017)
6. Box, G.E., Jenkins, G.M., Reinsel, G.C., Ljung, G.M.: Time series analysis: forecasting and control. Wiley, New York (2015)
7. Carbonneau, R., Laframboise, K., Vahidov, R.M.: Application of machine learning techniques for supply chain demand forecasting. Eur. J. Oper. Res. **184**(3), 1140–1154 (2008)
8. Chen, T., et al.: TADA: trend alignment with dual-attention multi-task recurrent neural networks for sales prediction. In: ICDM, pp. 49–58 (2018)
9. Chen, Y., Jin, X., Feng, J., Yan, S.: Training group orthogonal neural networks with privileged information. In: IJCAI, pp. 1532–1538 (2017)
10. Chollet, F., et al.: Keras (2015). https://keras.io
11. Fan, C., et al.: Multi-horizon time series forecasting with temporal attention learning. In: KDD, pp. 2527–2535 (2019)
12. Hoffman, J., Gupta, S., Darrell, T.: Learning with side information through modality hallucination. In: CVPR, pp. 826–834 (2016)

13. Kingma, D.P., Ba, J.: Adam: a method for stochastic optimization. In: ICLR (2015)
14. Koprinska, I., Wu, D., Wang, Z.: Convolutional neural networks for energy time series forecasting. In: IJCNN, pp. 1–8 (2018)
15. Ładyżyński, P., Żbikowski, K., Grzegorzewski, P.: Stock trading with random forests, trend detection tests and force index volume indicators. In: Rutkowski, L., Korytkowski, M., Scherer, R., Tadeusiewicz, R., Zadeh, L.A., Zurada, J.M. (eds.) ICAISC 2013. LNCS (LNAI), vol. 7895, pp. 441–452. Springer, Heidelberg (2013). https://doi.org/10.1007/978-3-642-38610-7_41
16. Lai, G., Chang, W., Yang, Y., Liu, H.: Modeling long- and short-term temporal patterns with deep neural networks. In: SIGIR, pp. 95–104 (2018)
17. Lambert, J., Sener, O., Savarese, S.: Deep learning under privileged information using heteroscedastic dropout. In: CVPR, pp. 8886–8895 (2018)
18. Lopez-Paz, D., Bottou, L., Schölkopf, B., Vapnik, V.: Unifying distillation and privileged information. In: ICLR (2016)
19. Oord, A.V.D., et al.: Wavenet: a generative model for raw audio. arXiv preprint arXiv:1609.03499 (2016)
20. Qi, Y., Li, C., Deng, H., Cai, M., Qi, Y., Deng, Y.: A deep neural framework for sales forecasting in e-commerce. In: CIKM, pp. 299–308 (2019)
21. Qin, Y., Song, D., Chen, H., Cheng, W., Jiang, G., Cottrell, G.W.: A dual-stage attention-based recurrent neural network for time series prediction. In: IJCAI, pp. 2627–2633 (2017)
22. Ristanoski, G., Liu, W., Bailey, J.: Time series forecasting using distribution enhanced linear regression. In: Pei, J., Tseng, V.S., Cao, L., Motoda, H., Xu, G. (eds.) PAKDD 2013. LNCS (LNAI), vol. 7818, pp. 484–495. Springer, Heidelberg (2013). https://doi.org/10.1007/978-3-642-37453-1_40
23. Salinas, D., Flunkert, V., Gasthaus, J., Januschowski, T.: DeepAR: probabilistic forecasting with autoregressive recurrent networks. Int. J. Forecast. **36**(3), 1181–1191 (2019)
24. Shen, Z., Zhang, Y., Lu, J., Xu, J., Xiao, G.: SeriesNet: a generative time series forecasting model. In: IJCNN, pp. 1–8 (2018)
25. Temporal pattern attention for multivariate time series forecasting: Shih, S.-Y., Sun, F.-K., Lee, H.-y. Mach. Learn. **108**, 1421–1441 (2019). https://doi.org/10.1007/s10994-019-05815-0
26. Srivastava, N., Hinton, G.E., Krizhevsky, A., Sutskever, I., Salakhutdinov, R.: Dropout: a simple way to prevent neural networks from overfitting. J. Mach. Learn. Res. **15**(1), 1929–1958 (2014)
27. Stein, R.M.: Benchmarking default prediction models: pitfalls and remedies in model validation. Moody's KMV, New York, 20305 (2002)
28. Sutskever, I., Vinyals, O., Le, Q.V.: Sequence to sequence learning with neural networks. In: NeurIPS, pp. 3104–3112 (2014)
29. Vairagade, N., Logofatu, D., Leon, F., Muharemi, F.: Demand forecasting using random forest and artificial neural network for supply chain management. In: Nguyen, N.T., Chbeir, R., Exposito, E., Aniorté, P., Trawiński, B. (eds.) ICCCI 2019. LNCS (LNAI), vol. 11683, pp. 328–339. Springer, Cham (2019). https://doi.org/10.1007/978-3-030-28377-3_27
30. Vapnik, V., Vashist, A.: A new learning paradigm: learning using privileged information. Neural Netw. **22**(5–6), 544–557 (2009)
31. Wen, R., Torkkola, K., Narayanaswamy, B., Madeka, D.: A multi-horizon quantile recurrent forecaster. arXiv preprint arXiv:1711.11053 (2017)
32. Wu, X., Shi, B., Dong, Y., Huang, C., Faust, L., Chawla, N.V.: RESTful: resolution-aware forecasting of behavioral time series data. In: CIKM, pp. 1073–1082 (2018)

GANNSTER: Graph-Augmented Neural Network Spatio-Temporal Reasoner for Traffic Forecasting

Carlos Salort Sánchez[1,2(✉)], Alexander Wieder[1], Paolo Sottovia[1],
Stefano Bortoli[1], Jan Baumbach[3], and Cristian Axenie[1]

[1] Huawei German Research Center, Munich, Germany
{carlos.salort,cristian.axenie}@huawei.com
[2] Technical University of Munich, Munich, Germany
[3] Chair of Experimental Bioinformatics, TUM School of Life Sciences
Weihenstephan, Freising, Germany

Abstract. Traffic forecast is a problem of high interest due to its impact on mobility and inherent socio-economic aspects of people's lives. Particularly for adaptive traffic light systems, the ability to predict traffic throughput in intersections enables fast adaptation, thus reducing traffic jams. In this work, we propose a novel approach for traffic forecasting, termed Graph Augmented Neural Network Spatio-TEmporal Reasoner (GANNSTER), which fuses spatial information, given by the traffic network topology, with temporal reasoning and learning capabilities of recurrent neural networks. Our modelling contribution is supplemented by the public release of a novel real-world dataset containing urban traffic throughput in intersections. We comparatively evaluate GANNSTER against state-of-the-art models for traffic forecast and demonstrate its superior performance.

Keywords: Deep learning · Graph neural network · Traffic forecast

1 Introduction

Traffic congestion resulting from growing traffic volumes in urban areas has a major impact on life, ranging from socio-economic to environmental aspects, such as air pollution, commute time, and waste of energy. One key method for reducing congestion is to optimize the traffic light control accordingly based on the current traffic situation, but also the expected traffic in the near future. For instance, accurate traffic flow forecasts can be used to improve traffic light control, therefore reducing the formation of jams or minimizing their effects.

Forecasting traffic flow, however, constitutes a complex problem. Traffic depends on a large variety of factors, for instance, the length of the traffic light phases, the type of vehicles, the driver behaviour, and the variety of weather conditions. Additionally, datasets present high variability between consecutive

© Springer Nature Switzerland AG 2020
V. Lemaire et al. (Eds.): AALTD 2020, LNAI 12588, pp. 63–76, 2020.
https://doi.org/10.1007/978-3-030-65742-0_5

readings from the road sensors. Such variability of consecutive measurements is illustrated with an example in Fig. 1. Said measurements may be subject to noise due to unforeseen external conditions (e.g. untrimmed trees, poor visibility at night, vandalism, etc.).

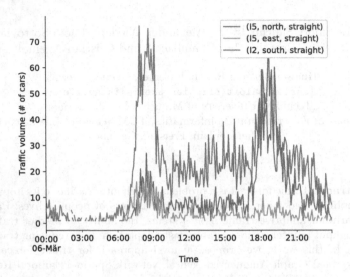

Fig. 1. Example for car throughput in an urban intersection. Data is aggregated every five minutes. Each line represents 24 h of data for one traffic sensor.

A variety of different techniques for traffic forecasting have been proposed, ranging from statistical methods [1] to Deep Learning models [13,18]. Deep Learning (DL) methods have proved their potential in learning from large amounts of data. For instance, DL excels in describing sequences of temporally dependent values [4,12], hierarchical visual processing [15] and data generation [24]. While these studies present promising results in restricted scenarios, they fail to capture the spatio-temporal relationships in the data. In an attempt to introduce this dimension into the model, there has been a large amount of work addressing methods to embed graph structures into neural networks [2,7,10,14,28,33]. Some of this work has been applied in the traffic domain, but most of it focuses on forecasting the average speed of vehicles driving through detectors on highways rather than throughput of urban traffic.

In this work, we address the challenge of urban traffic flow forecasting, and introduce *GANNSTER: Graph Augmented Neural Network Spatio-TEmporal Reasoner*, a novel deep learning model and system that exploits both temporal and spatial information through embedded graphs for predicting traffic flow. The primary purpose of GANNSTER is to provide traffic flow predictions to be used for traffic light logic optimization. Since traffic light logic optimization requires predictions at most about one hour ahead, GANNSTER focuses on this short-term prediction horizon.

Our main contributions are:

- GANNSTER, a neural network-based system, that embeds and exploits the temporal and spatial relations in a road network for throughput prediction across intersections.
- The comparative evaluation of GANNSTER against relevant state-of-the-art models on a novel real-world dataset. Our evaluation results show that our approach generally yields higher accuracy than the other evaluated methods.
- A new real-world dataset, MUSTARD-S (Multi-cross Urban Signalized Traffic Aggregated Region Dataset - Small), which contains road traffic data recorded over 55 days and 6 intersections.

We start by describing related work employing recurrent neural networks, graph neural networks and other methods for traffic forecasting in Sect. 2. Following, in Sect. 3, we introduce GANNSTER. The experiments, along with their methodology, are introduced in Sect. 4. We present and discuss the evaluation and experimental results in Sect. 5. Finally, Sect. 6 concludes the paper and discusses opportunities for future work.

2   Related Work

The proposed system taps into efficient solutions for traffic forecasting and explores how the new breed of graph neural networks can tackle the inherent dynamics of such a complex process. In the following, we provide an overview of key state-of-the-art approaches.

2.1   Traffic Forecasting

Most of the traffic forecasting approaches use statistical methods [1,9]. While these types of models work well on small datasets, the increasing number of traffic sensors, data quantities and heterogeneity, and computational power has made them obsolete. Recent approaches use machine learning techniques for detecting non-linear relations among the traffic variables. For instance, works using Support Vector Regression [25] tend to outperform statistical methods in terms of accuracy by learning a linear function in the space induced by a non-linear kernel which corresponds to a non-linear function in the original space. Following the trend from the machine learning community, researchers turned towards DL for traffic forecasting [22,26]. While outperforming both statistical and machine learning methods, base DL methods fail to exploit temporal and spatial information. With the rise of recurrent neural networks, and particularly Long Short-Term Memory (LSTM) and Gated Recurrent Units (GRU) [4,12], researchers added the temporal dependency to the Eq. [20,31]. Yet, such systems still fail to capture the spatial information. In an attempt to add spatial information, some models incorporated convolutional neural networks [19,30,32], with the downside of not being able to accurately represent the road network topology.

2.2   Graph Neural Networks

Recent trends demonstrated an increasing interest in combining DL techniques and graphs. Whereas most data problems lie in a Euclidean space, that is not the case for graph data. Therefore some of the data assumptions do not hold (e.g. hierarchical representation, flatness, flexible operations with fewer dimensions).

To bypass these difficulties, some approaches included the use of graph-structured spatial information, such as the PATCHY-SAN that extracted locally connected regions from graphs using learned feature representations competitive with state-of-the-art graph kernels [21]. GraphSAGE is another approach that employs inductive learning to leverage graph node attribute information to efficiently generate representations of previously unseen data [10]. Another remarkable approach leverages Diffusion Convolution Networks that introduce a novel diffusion-convolution operation and diffusion-based representations that can be learned from graph-structured data and used as an effective basis for node classification [17]. Finally, Attention Based Methods stand out, particularly self-attention mechanisms, that relate different positions of a single sequence to compute a representation of the same sequence [27]. Among these, the method that has gained more popularity relies on generalizing convolutional neural networks [16] using spectral graph theory. This idea was first introduced in [3], with Graph Convolutional Networks (GCN). Several works [6,11,14] have been built on top of these principles using different approximations from spectral graph theory. These methods have been successfully applied to a growing set of problems, such as link prediction in optimizing networks [14], and representing three-dimensional protein structures [7]. A more detailed description of general GCN can be found in [28,33].

2.3   Graph Neural Networks for Traffic Forecasting

With the evolution of graph-based neural networks, a new opportunity to tackle traffic forecast problems arose. The work in [29] proposes Spatio-Temporal GCN (STGCN) to combine graph convolutions and gated temporal convolutions for extracting the most relevant spatial and temporal features coherently. By equipping a neural network with attention mechanisms, the work in [8] enabled focusing on a subset of inputs and features. Basically, by computing masks used to multiply features, the work proposes an attention mechanism with three independent temporal components, namely recent data, daily data, and weekly data, fused to generate the final traffic forecast. The work in [17] introduces Diffusion Convolutional Recurrent Neural Networks (DCRNN), a model employing bidirectional random walks on a graph to learn its spatial dependencies, and an encoder-decoder with scheduled sampling architecture to detect the temporal dependencies. Our method differs structurally from this approach, as we train the model directly, and propose an alternative convolution-based approach. Another important aspect is the fact that DCRNN uses a weighted graph based on the distance between sensors, which we avoid to reduce the amount of information needed about the network. Finally, the work in [5] proposes Traffic Graph Convolutional LSTM (TGC-LSTM), a model based on LSTM, that uses Free-Flow Reachability (FFR) matrices in the graph convolution to provide extra information to the model. Similarly to our work, TGC-LSTM uses k-walks matrices, but without considering previous temporal values. Furthermore, our model does not depend on any extra information other than the adjacency matrix.

3 GANNSTER

In this section, we detail the structure of GANNSTER and the graph encoding of the road network to exploit spatio-temporal dependencies for traffic predictions. We start by introducing the road graph and definitions used throughout this section.

3.1 Road Graph

The road network structure is crucial for producing accurate traffic forecasts. For instance, knowing that a road segment is unidirectional often allows reasonably accurate predictions for the next intersection reached by a vehicle driving on that road. While this is intuitive for humans, such properties need to be carefully encoded into the model to enable forecasts.

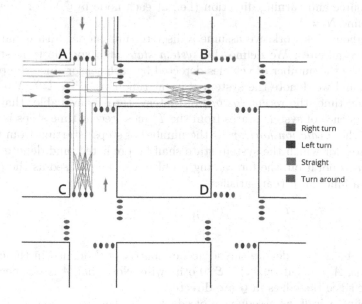

Fig. 2. Possible paths for a car entering intersection A. There are four possible directions, and four possible turns. As an example, a car entering (A, north, left) has four possible destinations: (B, west, {any direction}).

Figure 2 depicts an example of the graph derived from a map with four intersections. For the sake of readability, possible paths are only plotted for intersection A. Each intersection can be entered from four possible directions (north, east, south, and west), and can be left into four directions (by turning left, right, going straight, and turning around). Note that the topology in the depicted example was solely chosen for clarity, and other road structures can be encoded similarly. In the following subsection, we formally define the construction of the road graph.

3.2 Definitions

We define the road graph as a directed graph $\mathcal{G} = (\mathcal{V}, \mathcal{E})$, where each vertex

$$v = (\text{intersection, source direction, turn}) \in \mathcal{V} \tag{1}$$

represents one concrete possibility for traversing an intersection (coming from a specific direction and taking a specific turn). We define that a directed edge

$$e = (v_{origin}, v_{destination}) \in \mathcal{E} \tag{2}$$

exists if and only if an intersection can be traversed as specified by $v_{destination}$ directly after traversing the same or a different intersection as specified by v_{origin}. We assume that sensors are installed at intersections counting vehicles for each possible source and turning direction (i.e., at each node in $\mathcal{G}$.). For brevity, we further define $N = |\mathcal{V}|$.

Throughout this work, we assume a discrete time model with equally-sized steps of five minutes. We define the *system state* at a specific time step t to consist of the the number of vehicles detected by each sensor since the previous time step, and we denote the system state at time t as $x^{(t)} \in \mathbb{R}^N$. We assume that, at any time, the recent *history* of system states is available, that is, the ordered sequence of system states from the T' most recent time steps is known. We define the *prediction horizon* as the number of steps (starting from the last known state) for which the system state shall be predicted, and denote it as T. Based on this notation, the forecasting problem can be phrased as the problem of finding a function h that satisfies

$$[x^{(t-T'+1)}, \ldots, x^{(t)}; \mathcal{G}] \xrightarrow{h} [x^{(t+1)}, \ldots, x^{(t+T)}]. \tag{3}$$

We let $A \in \mathbb{R}^{N \times N}$ denote the adjacency matrix of $\mathcal{G}$ defined in the common way, that is, $A_{i,j} = 1$ if $(v_i, v_j) \in \mathcal{E}$, 0 otherwise. Note that A is not necessarily symmetric since the edges in $\mathcal{G}$ are directed.

We define a *walk* as a sequence of edges $[e_1 = (v_0, v_1), \ldots, e_i = (v_{i-1}, v_i)]$, which connects a sequence of vertices in the graph. With M^k denoting the k'th power of a matrix M, given the adjacency matrix A of $\mathcal{G}$, the matrix A^k represents the number of possible walks of degree k. That is, $A_{i,j}^k$ represents the number of walks from vertex v_i to v_j with length k. In a road traffic graph, this can be interpreted as the multitude of nodes that can be reached in k time steps by a vehicle originally detected in vertex v_i.

We define the *k-walk matrix* as

$$\hat{A}_{i,j}^k = min(A_{i,j}^k, 1), \tag{4}$$

such that $\hat{A}_{i,j}^k = 1$ if there is at least one k-degree walk from v_i to v_j, and $\hat{A}_{i,j}^k = 0$ otherwise. Each row and column of this matrix represents one vertex in the graph, and the matrix represents the final vertex (columns) where a vehicle can arrive starting from the initial vertex (row) in k steps. After multiplying $\hat{A}^k$ with the system state, we obtain $\hat{A}^k x^{t-k}$, a vector representing the maximum amount of cars that can arrive at a particular node from any node in the network in k steps.

We let $D^k \in \mathbb{R}^{N \times N}$ denote the *degree matrix* of $\hat{A}^k$: $D_{ii}^k = \sum_{j=1}^{N} \hat{A}_{i,j}^k$. This diagonal matrix represents the number of edges that can be reached in exactly k steps starting from the vertex v_i. The inverse of the degree matrix is represented by D^{-k}.

The GANNSTER model utilizes a similar approach to the *Graph Convolution* operation defined in [11, 14], using the adjacency matrix as $\hat{A} x^{(t)}$ to extract local information from previous steps.

3.3   GANNSTER Model

GANNSTER incorporates both temporal and spatial information by leveraging a combination of Graph Convolutions and Recurrent Neural Networks (RNN).

Temporal Information. GANNSTER utilizes RNNs, an established type of DL structures designed for use with temporal data. GANNSTER is agnostic to concrete type of RNN, and in this work, we instantiate GANNSTER in combination with LSTM and GRU. RNNs are well-suited for processing sequence data for predictions but suffer from short-term memory. LSTMs and GRUs mitigate short-term memory using gates that regulate the flow of information flowing through the sequence chain. In addition to the temporal traffic information, in either case, we augment the input vector for the t-step with additional information from the graph representing the road network.

Spatio-Temporal Information. Road topology contains rich implicit information (e.g. adjacency, connectivity, directions). Our objective is to incorporate this spatio-temporal information into the RNN components of GANNSTER.

In Fig. 3 we can see a vehicle, currently positioned in (A, west, straight), at time t. Let's assume that, in one timestep, it can move one intersection. Then, due to the road topology, we know that the vehicle will be in (C, west, any direction) at time $t + 1$. Furthermore, at time $t + 2$ the vehicle can be in (A,

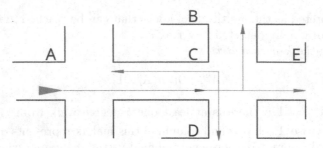

Fig. 3. Vehicle located in (A, west, straight), and possible paths in one timestep (blue) and in two timesteps (red). (Color figure online)

east, any), (B, south, any), (D, north, any) or (E, west, any). This information can be used by the model to improve forecast accuracy.

We will use the matrix $\hat{A}^k$ to incorporate spatio-temporal information. As it represents the possible k-walks, when computing the product $\hat{A}^k x^{(t-k)}$, we obtain the number of vehicles from k timesteps ago and k hops away from i at position i. This represents a rich new source of information that constitutes the base for our model.

GANNSTER Network. GANNSTER embeds the graph structure along with the temporal information into the model. We define the parameter K as the number of past steps that will be considered in the model.

We define a GANNSTER vector as

$$\text{GANNSTER}^t = \overset{K}{\underset{k=0}{\big\|}} (D^{-k} \hat{A}^k x^{(t-k)}) \tag{5}$$

where $\|$ represents vector concatenation. The vector described in Eq. 5 will be the input of the RNN. In cases where $(t - k) < (t - T' + 1)$, i.e. the input information for the model is not available because it is too old, we use $x^{(t-k)} = 0$. We use D^{-k} to normalize the number of cars in previous steps.

Figure 4 shows the architecture used by GANNSTER, when used jointly with LSTM. As explained before, other RNN structures can be used. We use many to many sequence prediction. The main difference with plain RNN architectures is the addition of the spatio-temporal information as input. Please note that, for the case $K = 0$, GANNSTERLSTM becomes a normal LSTM model. Analogous to other RNN-based systems, we can stack L blocks. We explored large scale structures and added a dropout layer between blocks. We only use GANNSTER vectors in the first block. In posterior blocks, the hidden space dimension does not necessarily match the input space dimension and therefore the k-walk matrix loses its meaning.

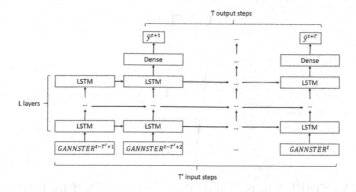

Fig. 4. GANNSTER architecture.

4 Experimental Evaluation

We conducted an experimental evaluation of GANNSTER to assess the performance in terms of forecasting accuracy. For comparison, we included state-of-the-art models that incorporate topological information about the road network encoded as graph, as well as simpler baseline models in our evaluation. In contrast to GANNSTER and the state-of-the-art methods, the baseline methods are oblivious to the structure of the road network, and hence, the comparison to them may indicate the performance benefit stemming from the additional topological information. For evaluating the performance in an urban setting, we introduce a novel dataset, MUSTARD-S (Multi-cross Urban Signalized Traffic Aggregated Region Dataset - Small), which we describe next.

4.1 MUSTARD-S

We present MUSTARD-S (Multi-cross Urban Signalized Traffic Aggregated Region Dataset - Small), a dataset consisting of 55 days of traffic throughput at six intersections in a city in China. We are working to increase the size of the dataset, and will be made public once available. The road network underlying this dataset is depicted in Fig. 5.

4.2 Experimental Settings

For training, we use a 80/10/10 train/validation/test split. Due to the time dependency of the data samples, we chose a sequential split. For all models, we train for up to 400 epochs, with an initial learning rate of 0.0001, and Mean Squared Error (MSE) as a loss function. We have a patience mechanism for updating the learning rate. Once the validation error does not improve by at least 0.00001 for 10 iterations, we decrease the learning rate by a factor of 10, resort to the iteration that achieved the best accuracy in validation and resume training from that state on. We stop training after the learning rate was updated

Fig. 5. MUSTARD-S map. Named intersections are considered in the study.

twice, or the epoch limit is reached. We normalize the dataset using Z-score. History size is one hour of data (12 data points) for all models. We consider a prediction horizon of 5, 15, 30, 45 and 60 minutes, respectively, into the future (i.e. 1, 3, 6, 9 and 12 values).

We considered the following models in our evaluation:

- Naïve Baseline. The prediction is the last value observed, regardless of the prediction horizon.
- DNN. It is a one layer dense neural network. The first layer has $N \cdot T'$ nodes, output layer has $N \cdot T$, where N is the number of vertices $|\mathcal{V}|$ in the road graph, T' is the history size, and T is the number of steps to predict.
- LSTM, GRU. Vanilla three layers stacked LSTM and GRU, with a hidden state of 128 nodes, dropout of 0.2, as implemented in PyTorch.
- TGC-LSTM. We use most of the same parameters as in [5]. We use $K = 3$, i.e. 3 steps behind. For the FFR matrix we use $\hat{A}^k$ as a proxy.
- GANNSTER-LSTM, GANNSTER-GRU, our proposed models, in LSTM and GRU flavours. Implemented using two layers stacked, with a hidden dimension of 128, and dropout of 0.2, and using the adjacency matrix for the GANNSTER vectors. We use $K = 3$, that is, 3 steps behind, as a sufficient history intake.

All models have been implemented using PyTorch [23]. Source code available at https://github.com/csalort/GANNSTER. The metrics used for forecast comparison are Mean Absolute Percentage Error (MAPE), Mean Absolute Error (MAE) and Root Mean Square Error (RMSE) for n samples, given ground truth y and prediction $\hat{y}$:

$$MAE = \frac{1}{n} \sum_{i=1}^{n} |y - \hat{y}| \tag{6}$$

$$MAPE = \frac{1}{n} \sum_{i=1}^{n} \left| \frac{y - \hat{y}}{y} \right| \cdot 100 \tag{7}$$

$$RMSE = \sqrt{\frac{1}{n} \sum_{i=1}^{n} (y - \hat{y})^2} \tag{8}$$

All experiments were conducted on a KunLun Mission Critical Server with 768 cores (equipped with Intel(R) Xeon(R) CPU E7-8890 v4 @ 2.20 GHz) and 12 TB of RAM.

5   Results and Discussion

Table 1 presents the results of the forecast experiments on the MUSTARD-S dataset, given as error between ground truth and prediction. First, we can observe that the forecasting accuracy generally drops as the prediction horizon is widened. We can also see that all the models have relatively poor performance, especially regarding to MAPE. This confirms our hypothesis that the high variability of the measurements create a challenging forecast environment.

Table 1. MUSTARD-S results. Best performance highlighted in bold.

		GANNSTER		TGC-LSTM	LSTM	GRU	DNN	Naïve baseline
		GRU	LSTM					
5 min	MAE	**2.346**	2.373	3.340	2.628	2.624	2.390	2.551
	MAPE	42.769	42.525	69.097	**41.460**	42.165	43.374	50.000
	RMSE	**4.055**	4.144	7.763	5.651	5.629	4.259	4.884
15 min	MAE	2.444	**2.407**	3.340	2.658	2.659	2.458	2.587
	MAPE	43.377	43.010	69.077	**41.781**	42.598	44.493	50.377
	RMSE	4.294	**4.246**	7.763	5.688	5.685	4.398	4.957
30 min	MAE	2.476	**2.443**	3.340	2.703	2.691	2.529	2.687
	MAPE	43.979	43.394	69.068	**42.485**	43.605	45.848	51.404
	RMSE	4.366	**4.345**	7.763	5.743	5.736	4.535	5.249
45 min	MAE	**2.567**	2.598	3.341	2.738	2.738	2.605	2.806
	MAPE	45.238	45.282	69.077	**43.008**	44.467	47.354	52.645
	RMSE	**4.634**	4.817	7.765	5.795	5.792	4.690	5.618
60 min	MAE	**2.619**	2.709	3.341	2.799	2.814	2.676	2.931
	MAPE	46.192	47.126	69.147	**44.753**	46.210	48.786	54.053
	RMSE	**4.825**	5.116	7.764	5.937	5.899	4.852	6.017

The Naïve baseline is one of the models performing worst. This type of model is unable to adapt to the variability of the measurements. Similar results can be observed for the DNN. While both models yield low MAPE results, they perform above average in the remaining metrics. This may be caused by an overfit during hours of low traffic, generating an overly low prediction model. On the contrary, RNN performs much better in MAPE. RNN can better adapt to traffic peaks and have some of the best scores in MAPE. Interestingly, the LSTM performs best in terms of MAPE. This is because MAPE results in a disproportionately high error in case of relatively small (true and predicted) traffic volumes. LSTM is similar to real values when there is not much traffic, but when the number of cars increases it stops performing so well. TGC-LSTM, the state-of-the-art model, performs quite poorly. Our hypothesis is that using $\hat{A}$ as a proxy for the FFR matrix hurts the model. It performs worse than all the baselines, thus making it unsuitable for the properties of the dataset. Our models, GANNSTERGRU and GANNSTERLSTM, are the best performers in two out of the three metrics, and rank second in the remaining. Moreover, the accuracy improvements of GANNSTERGRU and GANNSTERLSTM over GRU and LSTM, respectively, can be attributed to the incorporation of the spatio-temporal information into the model. GANNSTER-models improve the forecast with respect to both baselines and state-of-the-art, therefore being the best suited model for the problem at hand.

6   Conclusion and Future Work

In this paper, we present GANNSTER, a graph-based RNN model designed to forecast road traffic. Our experimental evaluation compares GANNSTER with state-of-the-art methods and baselines on a real-world dataset, which has been made public. We demonstrate through a performance analysis that GANNSTER outperforms the state-of-the-art in traffic flow forecast.

Our future lines of research include the possibility to use GANNSTER on "hidden traffic metrics", by further exploiting the intrinsic spatio-temporal mechanisms at its core. A different line of research is to incorporate more long-term traffic dynamics into GANNSTER to support a prediction horizon of days, enabling additional use-cases, such as improved city planning. Finally, we aim to extend the traffic datasets we used for evaluation, covering a larger area over a longer time, thus exploring the dynamics and robustness of the system at scale.

Acknowledgements. We want to thank our colleagues from Huawei for the insight and expertise provided during the development and writing of the paper. We would also like to thank the multiple reviewers of the paper for their useful critiques, which allowed us to write a better paper and to improve future lines of research. Jan Baumbach is grateful for the funding from H2020 project FeatureCloud, under grant agreement No. 826078.

References

1. Ahmed, M.S., Cook, A.R.: Analysis of freeway traffic time-series data by using Box-Jenkins techniques. No. 722 (1979)

2. Battaglia, P., Pascanu, R., Lai, M., Rezende, D.J., et al.: Interaction networks for learning about objects, relations and physics. In: Advances in Neural Information Processing Systems, pp. 4502–4510 (2016)
3. Bruna, J., Zaremba, W., Szlam, A., LeCun, Y.: Spectral networks and locally connected networks on graphs. arXiv preprint arXiv:1312.6203 (2013)
4. Chung, J., Gulcehre, C., Cho, K., Bengio, Y.: Empirical evaluation of gated recurrent neural networks on sequence modeling. arXiv preprint arXiv:1412.3555 (2014)
5. Cui, Z., Henrickson, K., Ke, R., Wang, Y.: Traffic graph convolutional recurrent neural network: a deep learning framework for network-scale traffic learning and forecasting. IEEE Trans. Intell. Transp. Syst. **21**, 4883–4894 (2019)
6. Defferrard, M., Bresson, X., Vandergheynst, P.: Convolutional neural networks on graphs with fast localized spectral filtering. In: Advances in Neural Information Processing Systems, pp. 3844–3852 (2016)
7. Fout, A., Byrd, J., Shariat, B., Ben-Hur, A.: Protein interface prediction using graph convolutional networks. In: Advances in Neural Information Processing Systems, pp. 6530–6539 (2017)
8. Guo, S., Lin, Y., Feng, N., Song, C., Wan, H.: Attention based spatial-temporal graph convolutional networks for traffic flow forecasting. Proc. AAAI Conf. Artif. Intell. **33**, 922–929 (2019)
9. Hamed, M.M., Al-Masaeid, H.R., Said, Z.M.B.: Short-term prediction of traffic volume in urban arterials. J. Transp. Eng. **121**(3), 249–254 (1995)
10. Hamilton, W., Ying, Z., Leskovec, J.: Inductive representation learning on large graphs. In: Guyon, I., Luxburg, U.V., Bengio, S., Wallach, H., Fergus, R., Vishwanathan, S., Garnett, R. (eds.) Advances in Neural Information Processing Systems, vol. 30, pp. 1024–1034. Curran Associates, Inc. (2017). http://papers.nips.cc/paper/6703-inductive-representation-learning-on-large-graphs.pdf
11. Henaff, M., Bruna, J., LeCun, Y.: Deep convolutional networks on graph-structured data. arXiv preprint arXiv:1506.05163 (2015)
12. Hochreiter, S., Schmidhuber, J.: Long short-term memory. Neural Comput. **9**(8), 1735–1780 (1997)
13. Jia, Y., Wu, J., Du, Y.: Traffic speed prediction using deep learning method. In: IEEE 19th International Conference on Intelligent Transportation Systems (ITSC), pp. 1217–1222. IEEE (2016)
14. Kipf, T.N., Welling, M.: Semi-supervised classification with graph convolutional networks. arXiv preprint arXiv:1609.02907 (2016)
15. LeCun, Y., Bengio, Y., Hinton, G.: Deep learning. Nature **521**(7553), 436–444 (2015)
16. LeCun, Y., et al.: Backpropagation applied to handwritten zip code recognition. Neural Comput. **1**(4), 541–551 (1989)
17. Li, Y., Yu, R., Shahabi, C., Liu, Y.: Diffusion convolutional recurrent neural network: data-driven traffic forecasting. arXiv preprint arXiv:1707.01926 (2017)
18. Lv, Y., Duan, Y., Kang, W., Li, Z., Wang, F.Y.: Traffic flow prediction with big data: a deep learning approach. IEEE Trans. Intell. Transp. Syst. **16**(2), 865–873 (2014)
19. Ma, X., Dai, Z., He, Z., Ma, J., Wang, Y., Wang, Y.: Learning traffic as images: a deep convolutional neural network for large-scale transportation network speed prediction. Sensors **17**(4), 818 (2017)
20. Ma, X., Tao, Z., Wang, Y., Yu, H., Wang, Y.: Long short-term memory neural network for traffic speed prediction using remote microwave sensor data. Transp. Res. Part C Emerg. Technol. **54**, 187–197 (2015)

21. Niepert, M., Ahmed, M., Kutzkov, K.: Learning convolutional neural networks for graphs. In: International Conference on Machine Learning, pp. 2014–2023 (2016)
22. Park, D., Rilett, L.R.: Forecasting freeway link travel times with a multilayer feedforward neural network. Comput.-Aided Civil Infrast. Eng. **14**(5), 357–367 (1999)
23. Paszke, A., et al.: Pytorch: an imperative style, high-performance deep learning library. In: Advances in Neural Information Processing Systems, pp. 8024–8035 (2019)
24. Radford, A., Metz, L., Chintala, S.: Unsupervised representation learning with deep convolutional generative adversarial networks. arXiv preprint arXiv:1511.06434 (2015)
25. Smola, A.J., Schölkopf, B.: A tutorial on support vector regression. Stat. Comput. **14**(3), 199–222 (2004)
26. Van Lint, J., Hoogendoorn, S., van Zuylen, H.J.: Freeway travel time prediction with state-space neural networks: modeling state-space dynamics with recurrent neural networks. Transp. Res. Rec. **1811**(1), 30–39 (2002)
27. Veličković, P., Cucurull, G., Casanova, A., Romero, A., Lio, P., Bengio, Y.: Graph attention networks. arXiv preprint arXiv:1710.10903 (2017)
28. Wu, Z., Pan, S., Chen, F., Long, G., Zhang, C., Yu, P.S.: A comprehensive survey on graph neural networks. CoRR abs/1901.00596 (2019). http://arxiv.org/abs/1901.00596
29. Yu, B., Yin, H., Zhu, Z.: Spatio-temporal graph convolutional neural network: a deep learning framework for traffic forecasting. CoRR abs/1709.04875 (2017). http://arxiv.org/abs/1709.04875
30. Yu, H., Wu, Z., Wang, S., Wang, Y., Ma, X.: Spatiotemporal recurrent convolutional networks for traffic prediction in transportation networks. Sensors **17**(7), 1501 (2017)
31. Yu, R., Li, Y., Shahabi, C., Demiryurek, U., Liu, Y.: Deep learning: a generic approach for extreme condition traffic forecasting. In: Proceedings of the 2017 SIAM International Conference on Data Mining, pp. 777–785. SIAM (2017)
32. Zhang, J., Zheng, Y., Qi, D.: Deep spatio-temporal residual networks for citywide crowd flows prediction. In: Thirty-First AAAI Conference on Artificial Intelligence (2017)
33. Zhou, J., Cui, G., Zhang, Z., Yang, C., Liu, Z., Sun, M.: Graph neural networks: a review of methods and applications. CoRR abs/1812.08434 (2018). http://arxiv.org/abs/1812.08434

A Model-Agnostic Approach
to Quantifying the Informativeness
of Explanation Methods for Time
Series Classification

Thu Trang Nguyen[✉], Thach Le Nguyen[✉], and Georgiana Ifrim[✉]

School of Computer Science, University College Dublin, Dublin, Ireland
thu.nguyen@ucdconnect.ie, {thach.lenguyen,georgiana.ifrim}@ucd.ie

Abstract. In this paper we focus on explanation methods for time series classification. In particular, we aim to quantitatively assess and rank different explanation methods based on their informativeness. In many applications, it is important to understand which parts of the time series are informative for the classification decision. For example, while doing a physio exercise, the patient receives feedback on whether the execution is correct or not (classification), and if not, which parts of the motion are incorrect (explanation), so they can take remedial action. Comparing explanations is a non-trivial task. It is often unclear if the output presented by a given explanation method is at all informative (i.e., relevant for the classification task) and it is also unclear how to compare explanation methods side-by-side. While explaining classifiers for image data has received quite some attention, explanation methods for time series classification are less explored. We propose a model-agnostic approach for quantifying and comparing different saliency-based explanations for time series classification. We extract importance weights for each point in the time series based on learned classifier weights and use these weights to perturb specific parts of the time series and measure the impact on classification accuracy. By this perturbation, we show that explanations that actually highlight discriminative parts of the time series lead to significant changes in classification accuracy. This allows us to objectively quantify and rank different explanations. We provide a quantitative and qualitative analysis for a few well known UCR datasets.

Keywords: Time series classification · Explainable machine learning · Evaluation · Comparing explanations · Saliency maps

1 Introduction

In the last decade, machine learning systems have become more ubiquitous and highly integrated with our daily life due to the increased availability of personal computing and wearable devices. Machine learning methods, including those dealing with time series data, have grown in complexity, performance,

© Springer Nature Switzerland AG 2020
V. Lemaire et al. (Eds.): AALTD 2020, LNAI 12588, pp. 77–94, 2020.
https://doi.org/10.1007/978-3-030-65742-0_6

and impact. Among many applications [22,23], Time Series Classification (TSC) algorithms are more commonly used nowadays in human activity recognition [3] tasks, which often require explanations for certain critical decisions [6,14]. This explanation is usually presented in the form of a *saliency map* [1], highlighting the parts of the time series which are *informative* for the classification decision.

Recent efforts both in designing intrinsically explainable machine learning algorithms, as well as building post-hoc methods explaining black-box algorithms, have gained significant attention [20,24,28,31,36]; yet, these efforts present us with a new challenge: *How to assess and objectively compare such methods?* In other words, if two methods give different explanations, e.g., two different saliency maps, which method and explanation should we trust more? Assessing and comparing explanations is a non-trivial problem and requires a solution.

In this work, we consider explanation and its informativeness within a defined computational scope, in which a more informative explanation means a higher capability to influence classifiers to correctly identify a class. With this definition, we aim to objectively quantify and compare the informativeness of different explanations, hence alleviating the need for, or at least reducing some of the effort for, conducting user-studies which are very difficult to reproduce [9]. We focus on quantitatively evaluating explanation methods for the TSC task. In this paper, we only consider methods that produce explanations in the form of saliency maps. In particular, we introduce a model-agnostic methodology to quantitatively assess and rank different saliency maps based on a concept of informativeness which we define in this paper.

In our experiments, we consider three popular and recent saliency-based explanation methods representing two approaches for generating explanations (i.e., model internals from an *intrinsically* explainable model and *post-hoc* explanations) and two scopes of explanations (i.e., *global* explanation for the entire dataset and *local* explanation for the prediction on a specific test example). As illustrated in Fig. 1, such methods often produce significantly different explanations and subsequently call for a methodology and evaluation measure for comparison.

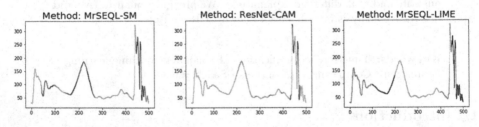

Fig. 1. Saliency map explanations for a motion time series obtained using different explanation methods. In this figure, the most discriminative parts are colored in deep red and the most non-discriminative parts are colored in deep blue. (Color figure online)

Our methodology stems from the idea that highly informative explanations correctly identify areas of the time series that are relevant for a classifier, thus perturbing such parts will result in a reduced capability of classifiers for making correct decisions. We focus on two scenarios in which the informativeness of explanation methods should be evaluated: when a *single explanation method* is presented and we want to know whether such method is actually informative, and when *multiple explanation methods* are presented and we wish to compare them.

The **evaluation of a single explanation method** compares the changes in classification performance under two settings: when the time series perturbation happens at either the discriminative and non-discriminative parts, as detected by the explanation method to be evaluated. If the method is informative, we expect that the accuracy will drop more significantly when the discriminative parts are perturbed. In contrast, for the **comparison of multiple explanation methods** we compare the classification performance only when the perturbation happens at the discriminative parts of the time series. The more informative method should trigger a more significant drop in accuracy. In both scenarios, we quantify the effect of change in performance by an **evaluation measure** which estimates the difference of the changes across multiple thresholds for identifying discriminative parts. We verify our experiment results with a sanity-check step, in which we visualize and compare the saliency maps for multiple examples of a dataset with known ground truth.

Our experiments show that explanations actually highlighting discriminative parts of the time series (i.e., that are more informative) lead to significant changes in classification accuracy, reflected by our proposed evaluation measure for quantifying this behaviour. While there is no one-size-fits-all ideal explanation method that perfectly highlights the discriminative parts in all TSC problems and datasets, our evaluation methodology provides a guideline to objectively evaluate any potential TSC saliency-based explanation methods for specific use cases, and safely reject those that fail both of the aforementioned steps.

Our main contributions are as follows:

1. We propose a new methodology and evaluation measure designed to enable us to objectively quantify and compare the informativeness of different explanation methods for the TSC task.
2. We empirically analyse our evaluation methodology with three representative explanation methods and three "referee" TSC algorithms.
3. We provide a discussion of the quantitative and qualitative assessment of various TSC explanation methods across several TSC benchmark datasets, and propose some directions for future work in this area.

2   Related Work

2.1   Time Series Classification

Although many TSC studies have been published in the past, very few of them focused on explainability. The list of TSC algorithms typically starts with the famous baselines 1NN-Euclidean and 1NN-DTW [16]; both are a combination of a nearest neighbour classifier and a distance measure. In most of the literature in this field, they are the benchmark classifiers due to their simplicity and accuracy. For this type of classifier, one can explain the classification decision on a time series by examining its nearest neighbour in the training data. However, we are not aware of any TSC studies that have investigated this prospect in depth.

Recent TSC papers have explored many other directions which include interval-based, shapelet-based, dictionary-based, and autocorrelation-based methods [4]. Nevertheless, only shapelet-based and dictionary-based classifiers in this group have shown the potential for explainability. Shapelet-based classifiers revolve around the concept of shapelets, segments of time series which can generalize or discriminate the classes. Examples of shapelet-based classifiers include Shapelet Transform [5] and Learning Shapelets [11]. It is theoretically possible to use shapelets as an explanation mechanism for these classifiers, but this was not considered in depth in previous studies, beyond a high-level qualitative discussion. On the other hand, dictionary-based classifiers have made significant breakthroughs with the introduction of SAX-VSM [29], BOSS [26], WEASEL [27], and Mr-SEQL [18]. The SAX-VSM work, although inferior to the latter in terms of accuracy, presented some attempts to explain the classifier by highlighting the highest-scored subsequences of the time series, which is a form of saliency mapping. Similar bids to explain the classification decision were made by SAX-VFSEQL [21] and its successor Mr-SEQL which are also classifiers from this group. Two other important families of TSC algorithms are deep learning, e.g., ResNet [15] and ensemble methods, e.g., HIVE-COTE [4]; they are generally well-known for being highly accurate. While not many attempts have been made to explain ensemble TSC methods, deep neural networks with convolutional layers can produce a saliency map explanation of the time series classification by using the Class Activation Map (CAM) method [36]. This option was explored in [35] and [15].

2.2   Explanation in Time Series Classification

Saliency Maps. Saliency mapping is a visualisation approach to highlight parts of a time series that are important for the TSC model in making a prediction. Such mappings are often produced by matching a time series with a *vector of weights* (w) using a color map. This vector of weights has a corresponding weight value for each data point in the time series. The saliency map (characterized by the vector of explanation weights) and the method to produce the vector of weights for the mapping, are hereafter respectively called TSC *explanation* and *explanation method*. Figure 2 (bottom right) shows an example saliency map in which the vector of explanation weights is matched to the original time series using a heatmap. The explanation weight is *non-negative* since its magnitude reflects the discriminative *power* of the associated data point in the time series.

In this work, we explore three TSC explanation methods using the concept of explanation seen as a vector of weights: **MrSEQL-SM, CAM**, and **LIME**. These methods represent three distinct approaches for producing saliency-based explanations in the form of highlighting the discriminative parts of a time series. We summarize the properties of these explanation methods in Table 1.

Table 1. Summary of TSC explanation methods properties.

Explanation method	Type	Model-specific	Explanation scope
MrSEQL-SM	Intrinsic	Yes	Global
CAM	Post-hoc	Yes	Local
LIME	Post-hoc	No	Local

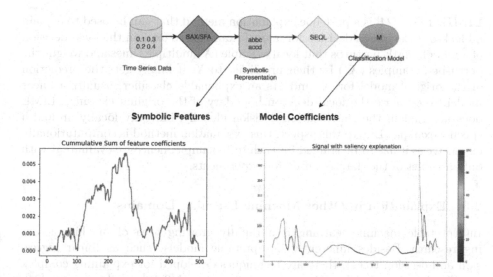

Fig. 2. The saliency map explanation MrSEQL-SM obtained from the MrSEQL linear classifier.

MrSEQL-SM. Mr-SEQL [18] is an efficient time series classification algorithm that is intrinsically explainable, since it learns a linear model for classification. The algorithm converts the numeric time series vector into strings, e.g., by using the SAX [19] transformation with varying parameters to create multiple symbolic representations of the time series. The symbolic representations are then used as input for SEQL [13], a sequence learning algorithm, to select the most discriminative subsequence features for training a classifier using logistic regression. The symbolic features combined with the classifier weights learned by logistic regression make this classification algorithm explainable (Fig. 2). For a time series, the explanation weight of each data point is the accumulated weight of the SAX features that it maps to. These weights can be mapped back to the original time series to create a saliency map to highlight the time series parts important for the classification decision. We call the saliency map explanation obtained this way, MrSEQL-SM. For using the weight vector from MrSEQL-SM, we take the absolute value of weights to obtain a vector of non-negative weights.

CAM. CAM [36] is a post-hoc explanation method commonly used to explain deep networks that have convolutional layers and a global average pooling (GAP) layer just before the final output layer. With this very specific architecture, the weights from the GAP layer can be used to reveal the parts of the time series that are important for the classifier to make a prediction. Thus, these weights are used to produce the saliency mapping of the weight vector to the original time series.

LIME. LIME [24] is a post-hoc explanation method that can be used to explain a black-box classifier's decision for a local example. To explain the local decision of a model, LIME perturbs that local example (X) multiple times and weighs the perturbed examples (X') by their proximity to X. It finally gets the prediction of the original model for X' and fits an explainable classifier, usually a linear model, to estimate the local decision boundary of the original classifier. LIME does not explain the classification decision globally, but only locally around a specific example. Due to this aspect, this explanation method is computationally expensive as it has to be trained for each test example, hence we evaluate it with only a subset of the datasets used for experiments.

2.3   Explanation in Other Machine Learning Domains

Interpretable machine learning is a rapidly growing area of machine learning research. Besides inherently interpretable models (such as linear regression and decision trees), there are techniques developed for explaining complex machine learning models, ranging from feature-based [2,10], local surrogate [24], to example-based explanations [25,34]. In the context of this work, we focus on studying explanation methods within the scope of saliency map explanation. Saliency maps were originally used in Computer Vision to highlight certain properties of the pixels in an image [30]. The success of black-box deep neural networks in image recognition tasks [12,17,33] paved the way for the growth of post-hoc explanation methods designed to explain deep learning models. Notable works of this family include Class Activation Map [36], Gradient-weighted Class Activation Map (Grad-CAM) [28] and Guided Backpropagation (GBP) [32]. This growing list of techniques to explain deep learning models poses the challenge of assessing the quality of these explanation methods. The work by [1] attempts to visually and statistically evaluate the quality of a few saliency-based explanations for deep learning models, by tracking the changes of the saliency maps when the model parameters and test labels are randomized. Interestingly, they show that some explanation methods provide unchanged explanations even when both the model parameters and the data are random.

3   Research Methods

In this section, we first describe the perturbation process we propose for evaluating the informativeness of a TSC explanation method. We then outline two perturbation approaches and finally introduce a novel measure to quantify and compare the informativeness of TSC explanation methods.

3.1   Explanation-Driven Perturbation

The goal of providing a TSC explanation is to focus on the discriminative parts of the time series. If the explanation is truly informative, it should point out those parts of the time series that are most relevant for the classification decision. Consequently, if we perturb these parts, then the time series will be harder to classify. The more informative the explanation, the higher the decrease in accuracy we expect, since we knock out the important information contained in the data, for making a classification decision. In this section we provide an approach for quantifying the informativeness of an explanation, by perturbing the data points, as guided by the explanation.

Discriminative weights are identified by a threshold k ($0 \leq k \leq 100$) that represent the $(100 - k)$-percentile of the non-negative weight vector (w) that explains a time series. This threshold allows us to focus on the highest magnitude weights in the vector, e.g., $k = 10$ means that we focus on the top 10% highest weights in the vector. With a specific value of k, the **discriminative parts** of the time series are those parts where w_t belongs to the $(100 - k)$-percentile discriminative weights. This part is important because the weight magnitude captures information about the discrimination power of the corresponding data point in the time series. Similarly, with the same threshold k, the **non-discriminative** parts of the time series are parts which have w_t in the k-percentile of the time series (e.g., for $k = 10$ these are the bottom 10% weights with lowest magnitude) (Fig. 3).

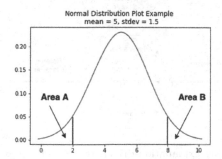

Fig. 3. Distribution of a hypothetical explanation weight vector with its non-discriminative weight area (Area A) and discriminative weight area (Area B).

We perturb a time series by adding Gaussian noise to its original signal. If the time series is represented by a vector x and the entire series is perturbed, the noisy time series would be represented by the new $x_{perturbed}$ vector

$$x_{perturbed} = x + \mathcal{N}(\mu, \sigma^2)$$

If a time series is normalized, the distribution for the Gaussian noise would be sampled from $\mathcal{N}(0, \sigma_1)$. The parameter σ_1 controls the magnitude of the noise.

In a similar fashion, the time series can also be selectively perturbed in accordance to a condition. In this case, we can perturb parts of the time series based on the corresponding weights in the explanation vector and keep the rest of the time series unchanged. With this logic of perturbing the time series (in accordance to a given weight vector), we selectively add noise to the time series as follows:

– *Type 1*: Perturbation applied only to discriminative region.
– *Type 2*: Perturbation applied only to non-discriminative region.

3.2 Method 1: Evaluating a Single Explanation Method

We propose an experiment to evaluate the informativeness of one explanation method. We aim to answer the question: *Is the explanation method truly informative?* In this experiment, we first build a time series classifier using the original, non-perturbed training time series. This classifier serves as the evaluation classifier for the explanation method, i.e., a *referee classifier*. In addition, we use the explanation method that we want to evaluate, to generate multiple versions of the test dataset, each corresponding to a value of the threshold k ($0 \leq k \leq 100$). For each value of k, we generate two perturbed test sets: one is only perturbed with *Type 1* noise, the other is only perturbed with *Type 2* noise. Using the referee classifier, we measure the accuracy in each perturbed test dataset. The entire process is summarized in Fig. 4.

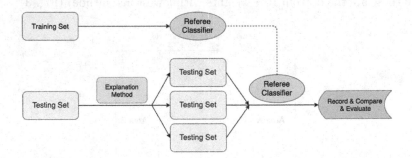

Fig. 4. Process of creating explanation-driven perturbed test sets and evaluating the explanation method using a referee classifier.

If the explanation method being evaluated is indeed informative, we expect that the perturbation of the discriminative parts (test datasets with *Type 1* noise) reduces the classifiers accuracy more than the perturbation of the non-discriminative parts (test datasets with *Type 2* noise).

3.3   Method 2: Comparing Multiple Explanation Methods

In contrast to the previous experiment, here we propose an experiment to compare multiple explanation methods by their informativeness. We follow the same process of creating noisy test sets as in Fig. 4, however, the perturbed test sets are now created differently. Instead of adding noise to both the discriminative and non-discriminative parts to create two different test sets for each k, we only add noise to the discriminative parts of the test time series. Since we have multiple explanation methods, at a same threshold k ($0 \le k \le 100$), we now have multiple versions of perturbed test datasets, each corresponding to a weight profile (i.e., explanation) obtained from one explanation method.

Among the evaluated explanation methods, a perturbation based on a more informative explanation should hurt the referee classifier more than the others. In Fig. 5, we hypothetically have two explanation methods with the *red* and *blue* lines representing the classification accuracy when test datasets are perturbed with either of the methods. Here, the explanation method controlling the perturbation of the test dataset with the resulting accuracy drawn in *red* is considered more informative, since perturbing the time series based on this explanation hurts the referee classifier more.

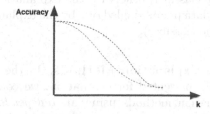

Fig. 5. Change of accuracy when the test set is perturbed with a threshold k.

As the vector of weights used as information to perturb the test dataset can be generated from *any* explanation method and independent of the referee classifier used to measure the change in accuracy, **Method 1** and **Method 2** are *model-agnostic* techniques to evaluate any TSC explanation method.

3.4   Informativeness of an Explanation: An Evaluation Measure

We quantify the informativeness of an explanation using the relationship between the accuracy of a referee classifier on test datasets perturbed at different levels k of noise. We calculate the impact of the explanation methods by estimating the area under the (explanation) curve described by accuracy at different perturbation levels k, using the trapezoidal rule. Since these values represents the reduction of the accuracy when noise is added to the time series, hereafter we call this metric *Explanation Loss* or *eLoss* for short. With this naming convention, one explanation method with lower *eLoss* will be considered better than another with higher *eLoss*.

$$eLoss = \frac{1}{2}k \sum_{i=1}^{t}(acc_{i-1} + acc_i)$$

where k denotes the values of each step normalized to 0–1 range; t denotes the number of steps $(t = \frac{100}{k})$; acc_i is the accuracy at step i. If we perturb the dataset with t steps, we will have a total of $t+1$ data points for accuracy scores. The step for $k = 0$ corresponds to the original test dataset, while the step for $k = 100$ corresponds to adding noise to the entire time series.

Evaluating a Single Explanation Method. The $eLoss$ can serve as a measure to evaluate the informativeness of one explanation method. In particular, we estimate the $eLoss$ of the accuracy curve produced by *Type 1* and *Type 2* noise. If the explanation method is informative, the *Type 1* $eLoss$ ($eLoss_1$) is expected to be less than *Type 2* $eLoss$ ($eLoss_2$). Alternatively, we can define this difference with Δ_{eLoss}:

$$\Delta_{eLoss} = eLoss_2 - eLoss_1.$$

If Δ_{eLoss} is positive, then the explanation method is computationally informative as captured by a referee classifier, otherwise the explanation method is deemed uninformative (i.e., the data points singled out by the explanation do not provide useful information to the classifier).

Comparing Multiple Explanation Methods. In the case where multiple explanation methods are presented for evaluation, we compare *Type 1 eLoss* ($eLoss_1$) for all explanation methods using an *independent referee classifier*. The explanation method that achieves the lowest $eLoss_1$ is the computationally most informative explanation method among the candidate methods.

4 Experiments

In this section, we present the results of applying our evaluation methodology using the following publicly available TSC datasets: CBF, Coffee, ECG200, Gun-Point from UCR [7] and the CMJ dataset[1]. TSC explanations for these datasets have been examined in depth by the previous works [15,18,29], hence they are suitable for demonstrating our approach. Table 2 summarizes these datasets.

We evaluate three TSC explanation methods: *MrSEQL-SM*, CAM based on ResNet (*ResNet-CAM*) and LIME based on the Mr-SEQL classifier (*Mr-SEQL-LIME*). We also train three referee classifiers, *Mr-SEQL* [18], *ROCKET* [8], and *WEASEL* [27], in order to computationally evaluate the usefulness of these explanation methods. Due to a high computational cost for LIME, we evaluate LIME only with the CMJ and GunPoint datasets. The code and settings for all our experiments are available at https://github.com/mlgig/explanation4tsc.

---

[1] Retrieved from: https://github.com/lnthach/Mr-SEQL/tree/master/data/CMJ.

Table 2. Summary of TSC datasets used to evaluate explanation methods.

Dataset	Train size	Test size	Length	Type	No. classes
CBF	30	900	128	Simulated	3
CMJ	419	179	500	Motion	3
Coffee	28	28	286	SPECTRO	2
ECG200	100	100	96	ECG	2
GunPoint	50	50	150	Motion	2

4.1 Experiment 1: Evaluation of a Single Explanation Method

Table 3 summarizes the results for the evaluation of the three explanation methods with the three referee classifiers over the five TSC datasets. We calculate the difference between *Type 2 eLoss* and *Type 1 eLoss* (Δ_{eLoss}) with the explanation-driven perturbation approach. We expect Δ_{eLoss} to be positive when the explanation method is informative.

Table 3. Summary of Δ_{eLoss} of three explanation methods on five different TSC problems. Positive values suggest the findings of the explanation method are informative according to the referee classifier. Negative values suggest otherwise.

Dataset	Explanation method	Referee classifier		
		Mr-SEQL	ROCKET	WEASEL
CBF	MrSEQL-SM	0.0001	0.002	0.0126
	ResNet-CAM	−0.0005	0.0007	0.0141
CMJ	MrSEQL-SM	0.0045	0.0709	0.1151
	ResNet-CAM	−0.0006	−0.0028	0.0106
	MrSEQL-LIME	0.0084	0.0475	0.0531
Coffee	MrSEQL-SM	0.0286	0.0	0.0
	ResNet-CAM	0.0179	0.0	0.0143
ECG200	MrSEQL-SM	0.033	−0.001	0.024
	ResNet-CAM	−0.011	−0.003	0.038
GunPoint	MrSEQL-SM	0.0026	0.1373	0.0273
	ResNet-CAM	0.0067	0.0967	−0.002
	MrSEQL-LIME	0.002	0.0714	0.0007

To visualize the difference between *Type 1 eLoss* and *Type 2 eLoss*, we also present this information in the form of the accuracy curve for the GunPoint dataset (Fig. 6) and the CMJ dataset (Fig. 7). In each of the figures, we draw the accuracy curve in the case when noise is added to the most discriminative parts (*Type 1*) and non-discriminative parts (*Type 2*). We note that if the *Type 1* curve

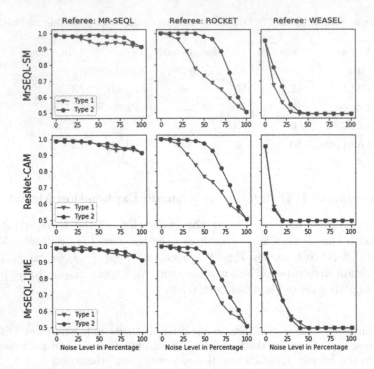

Fig. 6. Comparison of accuracy for *Type 1* (red) and *Type 2* (blue) perturbation for each explanation method and referee classifier for the GunPoint dataset. (Color figure online)

is below the *Type 2* curve, the explanation method is considered informative. If this trend is consistent across the referee classifiers, the evidence that the method is informative has more support. If we focus on evaluating the MrSEQL-SM explanation method for the GunPoint dataset, we observe that the *Type 1* curve is always below the *Type 2* curve for all three referee classifiers, thus we expect that this explanation method is informative. This information is consistent with the metric Δ_{eLoss} in Table 4, when Δ_{eLoss} is positive for all classifiers.

4.2   Comparison of Multiple Explanation Methods

In this experiment, we aim to compare the different explanation methods for a specific dataset. Instead of comparing the *eLoss* for the case when noise is added to the discriminative parts (*Type 1*) and non-discriminative parts (*Type 2*) of the time series for one explanation method, here we compare the *eLoss* for *Type 1* ($eLoss_1$) perturbation across different explanation methods. An explanation method is considered more informative if it has a smaller $eLoss_1$ for the same referee classifier.

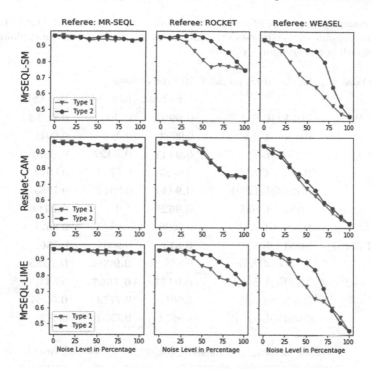

Fig. 7. Comparison of accuracy for *Type 1* (red) and *Type 2* (blue) perturbation for each explanation method and referee classifier for the CMJ dataset. (Color figure online)

We visualize this $eLoss_1$ in Fig. 8 in which the explanation curve of the three examined explanation methods are compared for the dataset CMJ (upper charts) and GunPoint (lower charts). We notice that the different in $eLoss_1$ is dependent on the referee classifier used to examine the change of the accuracy in the noisy test dataset. Given the same noisy datasets, the referee classifiers yield different classification accuracy. With the CMJ dataset, it is difficult to conclude which explanation method is most informative from Fig. 8, since the three lines are closely placed. This result is consistent with the comparison of $eLoss_1$ in Table 4. We can conclude that the three explanation methods are computationally similar in informativeness, although MrSEQL-SM is slightly more informative than the other two methods (its $eLoss_1$ is lowest for two referee classifiers).

4.3   Sanity Checks for Experiment Results

Although the evaluation measures show that one explanation method is more informative than another, we want to verify this conclusion by performing a sanity check step. In this step, we plot a few classification examples and their explanations by the methods evaluated previously. We choose to perform this

Table 4. Summary of $eLoss_1$ of three explanation methods on five different problems. Lower value (column-wise) suggests the explanation method is better in explaining the problem according to the referee classifier.

Dataset	Explanation method	Referee classifier		
		Mr-SEQL	ROCKET	WEASEL
CBF	MrSEQL-SM	**0.9991**	**0.9941**	**0.6018**
	ResNet-CAM	0.9993	0.9945	0.6041
CMJ	MrSEQL-SM	**0.9441**	**0.8422**	**0.6899**
	ResNet-CAM	0.9453	0.8735	0.6972
	MrSEQL-LIME	**0.9441**	0.8612	0.7385
Coffee	MrSEQL-SM	**0.9625**	1.0	**0.9786**
	ResNet-CAM	0.9696	1.0	0.9821
ECG200	MrSEQL-SM	**0.811**	0.9065	0.7565
	ResNet-CAM	0.838	**0.9035**	**0.7385**
GunPoint	MrSEQL-SM	**0.9477**	**0.7567**	0.543
	ResNet-CAM	0.961	0.7773	**0.5257**
	MrSEQL-LIME	0.9677	0.7953	0.573

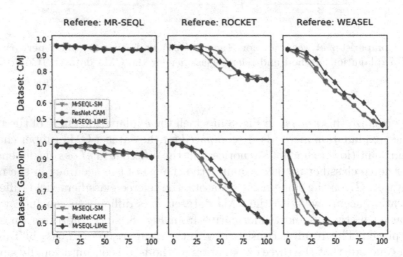

Fig. 8. Comparison of accuracy for *Type 1* perturbation based on three explanation methods (MrSEQL-SM, ResNet-CAM and Mr-SEQL-LIME) for GunPoint and CMJ datasets and three referee classifiers. Lower curve is better.

step with the CMJ dataset, for which the explanations are verified by a domain expert [18].

Figure 9 presents the saliency maps generated by three explanation methods for examples from the three motion classes in CMJ. Here we clearly see that these methods give different explanations. MrSEQL-SM seems to provide the

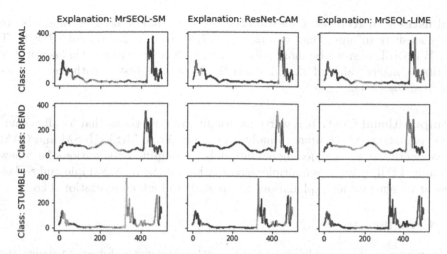

Fig. 9. Saliency maps produced by three explanation methods for example time series from the three classes of the CMJ dataset.

most informative/correct explanations that highlight the low, middle parts of the class NORMAL, the hump, middle parts of the class BEND, and the very high peak, middle parts of the class STUMBLE. MrSEQL-LIME gives a similar picture since it tries to explain the same classifier as MrSEQL-SM. ResNet-CAM does not clearly highlight similar parts in this dataset. This sanity check confirms the quantitative results in the previous experiments.

5 Discussion

In this section, we holistically interpret the experiment results with regard to informativeness and other perspectives. With the notion of informativeness, we set up the experiments based on an explanation-driven perturbation approach. This approach allows us to assess the contributing significance of the discriminative parts for a referee classifier. The results show that, with a given dataset, we are able to some extent evaluate and quantify the informativeness of different TSC explanation methods. There is scope though for further study of other perturbation approaches as well as the use of other referee classifiers in order to reach more significant differences in informativeness levels.

Stability of Explanation. Performing the experiment repeatedly, we notice that not all explanation methods provide consistent results. Methods that depend on certain level of randomization such as CAM (with randomized weight initialization) and LIME (with randomized local examples to estimate explanations) are generating slightly different explanations in different runs. For methods that are characterized by many hyperparameters like LIME, this stability of explanation is also dependent on these parameters, such as the number of local examples that it generates.

Robustness of Referee Classifier. We observe that some TSC methods are more sensitive to noise than others (Fig. 6, 7, 8). In our experiment, ROCKET and WEASEL seem to be more noise-sensitive than Mr-SEQL. This sensitivity results in higher value of Δ_{eLoss} when the method is tested with these noise-sensitive classifiers.

Computational Cost. It is worth mentioning that methods that locally generate explanations are computationally expensive. While MrSEQL-SM and CAM conveniently use the trained model internals to compute explanations for a new example, LIME generates multiple perturbations of the new example and reclassifies it to generate an explanation, which leads to high computational cost.

6   Conclusion

This work aims to provide an objective evaluation methodology to gauge the informativeness of explanation methods. Our experiment results show that it is feasible to quantitatively assess TSC explanation methods and the sanity checks visually confirm the experiment results. We envision that this technique is helpful when a user wants to assess an existing explanation method in the context of a given application, or wishes to evaluate different methods and opt for one that works best for a specific use case. In the scope of this work, we primarily evaluate three explanation methods which collectively represents different approaches to explain TSC decisions, though there are many other methods worth exploring. With the application of human activity recognition in mind, we believe that advancement in this area can potentially help many people who can thus conveniently access high quality technology to directly improve their lives.

Acknowledgments. This work was funded by Science Foundation Ireland through the SFI Centre for Research Training in Machine Learning (18/CRT/6183), the Insight Centre for Data Analytics (12/RC/2289_P2) and the VistaMilk SFI Research Centre (SFI/16/RC/3835).

References

1. Adebayo, J., Gilmer, J., Muelly, M., Goodfellow, I., Hardt, M., Kim, B.: Sanity checks for saliency maps. In: Proceedings of the 32nd International Conference on Neural Information Processing Systems, NIPS 2018, pp. 9525–9536. Curran Associates Inc., Red Hook (2018)
2. Apley, D.W., Zhu, J.: Visualizing the effects of predictor variables in black box supervised learning models (2016)
3. Avci, A., Bosch, S., Marin-Perianu, M., Marin-Perianu, R., Havinga, P.: Activity recognition using inertial sensing for healthcare, wellbeing and sports applications: a survey, pp. 167–176 (01 2010)
4. Bagnall, A., Lines, J., Bostrom, A., Large, J., Keogh, E.: The great time series classification bake off: a review and experimental evaluation of recent algorithmic advances. Data Mining and Knowledge Discovery, 1–55 (2016). https://doi.org/10.1007/s10618-016-0483-9

5. Bostrom, A., Bagnall, A.: Binary Shapelet transform for multiclass time series classification. In: Madria, S., Hara, T. (eds.) DaWaK 2015. LNCS, vol. 9263, pp. 257–269. Springer, Cham (2015). https://doi.org/10.1007/978-3-319-22729-0_20
6. Bostrom, N., Yudkowsky, E.: The ethics of artificial intelligence (2011)
7. Dau, H.A., et al.: Hexagon-ML: The UCR time series classification archive, October 2018. https://www.cs.ucr.edu/~eamonn/time_series_data_2018/
8. Dempster, A., Petitjean, F., Webb, G.I.: Rocket: exceptionally fast and accurate time series classification using random convolutional kernels (2019)
9. Doshi-Velez, F., Kim, B.: Towards a rigorous science of interpretable machine learning (2017)
10. Fisher, A., Rudin, C., Dominici, F.: All models are wrong, but many are useful: learning a variable's importance by studying an entire class of prediction models simultaneously (2018)
11. Grabocka, J., Schilling, N., Wistuba, M., Schmidt-Thieme, L.: Learning time-series Shapelets. In: Proceedings of the 20th ACM SIGKDD International Conference on Knowledge Discovery and Data Mining, KDD 2014, pp. 392–401. ACM, New York (2014). https://doi.org/10.1145/2623330.2623613
12. He, K., Zhang, X., Ren, S., Sun, J.: Deep residual learning for image recognition. CoRR abs/1512.03385 (2015). http://arxiv.org/abs/1512.03385
13. Ifrim, G., Wiuf, C.: Bounded coordinate-descent for biological sequence classification in high dimensional predictor space. In: Proceedings of the 17th ACM SIGKDD International Conference on Knowledge Discovery and Data Mining, KDD 2011, pp. 708–716. Association for Computing Machinery, New York (2011). https://doi.org/10.1145/2020408.2020519
14. Ismail Fawaz, H., Forestier, G., Weber, J., Idoumghar, L., Muller,P.A.: Accurate and interpretable evaluation of surgical skills from kinematic datausing fully convolutional neural networks. Int. J. Comput. Assist. Radiol. Surg. **14**(9), 1611–1617 (2019).https://doi.org/10.1007/s11548-019-02039-4
15. Ismail Fawaz, H., Forestier, G., Weber, J., Idoumghar, L., Muller,P.A.: Deep learning for time series classification: a review. Data Min. Knowl Disc. (2019). https://doi.org/10.1007/s10618-019-00619-1
16. Keogh, E., Ratanamahatana, C.A.: Exact indexing of dynamic time warping. Knowl. Inf. Syst. **7**(3), 358–386 (2005). https://doi.org/10.1007/s10115-004-0154-9
17. Krizhevsky, A., Sutskever, I., Hinton, G.E.: ImageNet classification with deep convolutional neural networks. In: Pereira, F., Burges, C.J.C., Bottou, L., Weinberger, K.Q. (eds.) Advances in Neural Information Processing Systems, vol. 25, pp. 1097–1105. Curran Associates, Inc. (2012). http://papers.nips.cc/paper/4824-imagenet-classification-with-deep-convolutional-neural-networks.pdf
18. Le Nguyen, T., Gsponer, S., Ilie, I., O'Reilly, M., Ifrim, G.: Interpretable time series classification using linear models and multi-resolution multi-domain symbolic representations. Data Min. Knowl. Disc. **33**(4), 1183–1222 (2019). https://doi.org/10.1007/s10618-019-00633-3
19. Lin, J., Keogh, E., Wei, L., Lonardi, S.: Experiencing sax: a novel symbolic representation of time series. Data Min. Knowl. Disc. **15**(2), 107–144 (2007). https://doi.org/10.1007/s10618-007-0064-z
20. Lundberg, S.M., Lee, S.I.: A unified approach to interpreting model predictions. In: Guyon, I., Luxburg, U.V., Bengio, S., Wallach, H., Fergus, R., Vishwanathan, S., Garnett, R. (eds.) Advances in Neural Information Processing Systems, vol. 30, pp. 4765–4774. Curran Associates, Inc. (2017). http://papers.nips.cc/paper/7062-a-unified-approach-to-interpreting-model-predictions.pdf

21. Nguyen, T.L., Gsponer, S., Ifrim, G.: Time series classification by sequence learning in all-subsequence space. In: IEEE 33rd International Conference on Data Engineering (ICDE), pp. 947–958, April 2017. https://doi.org/10.1109/ICDE.2017.142

22. Petitjean, F., Forestier, G., Webb, G.I., Nicholson, A.E., Chen, Y., Keogh, E.: Dynamic time warping averaging of time series allows faster and more accurate classification. In: IEEE International Conference on Data Mining, pp. 470–479 (2014)

23. Ramgopal, S., et al.: Seizure detection, seizure prediction, and closed-loop warning systems in epilepsy. Epilepsy Behav. E&B **37C**, 291–307 (2014). https://doi.org/10.1016/j.yebeh.2014.06.023

24. Ribeiro, M.T., Singh, S., Guestrin, C.: Why should I trust you?: explaining the predictions of any classifier. CoRR abs/1602.04938 (2016). http://arxiv.org/abs/1602.04938

25. Ribeiro, M.T., Singh, S., Guestrin, C.: Anchors: High-precision model-agnostic explanations. In: AAAI (2018)

26. Schäfer, P.: The boss is concerned with time series classification in the presence of noise. Data Min. Knowl. Discov. **29**(6), 1505–1530 (2015)

27. Schäfer, P., Leser, U.: Fast and accurate time series classification with weasel. In: Proceedings of the 2017 ACM on Conference on Information and Knowledge Management, CIKM 2017, pp. 637–646. ACM, New York (2017). https://doi.org/10.1145/3132847.3132980

28. Selvaraju, R.R., Das, A., Vedantam, R., Cogswell, M., Parikh, D., Batra, D.: Grad-CAM: why did you say that? visual explanations from deep networks via gradient-based localization. CoRR abs/1610.02391 (2016). http://arxiv.org/abs/1610.02391

29. Senin, P., Malinchik, S.: SAX-VSM: interpretable time series classification using sax and vector space model. In: IEEE 13th International Conference on Data Mining (ICDM), pp. 1175–1180, December 2013. https://doi.org/10.1109/ICDM.2013.52

30. Simonyan, K., Vedaldi, A., Zisserman, A.: Deep inside convolutional networks: visualising image classification models and saliency maps. Preprint, December 2013

31. Smilkov, D., Thorat, N., Kim, B., Viégas, F.B., Wattenberg, M.: Smoothgrad: removing noise by adding noise. CoRR abs/1706.03825 (2017), http://arxiv.org/abs/1706.03825

32. Springenberg, J., Dosovitskiy, A., Brox, T., Riedmiller, M.: Striving for simplicity: the all convolutional net. In: ICLR (workshop track) (2015). http://lmb.informatik.uni-freiburg.de/Publications/2015/DB15a

33. Szegedy, C., et al.: Going deeper with convolutions. In: IEEE Conference on Computer Vision and Pattern Recognition (CVPR), pp. 1–9 (2015)

34. Wachter, S., Mittelstadt, B.D., Russell, C.: Counterfactual explanations without opening the black box: automated decisions and the GDPR. CoRR abs/1711.00399 (2017). http://arxiv.org/abs/1711.00399

35. Wang, Z., Yan, W., Oates, T.: Time series classification from scratch with deep neural networks: a strong baseline. In: International Joint Conference on Neural Networks (IJCNN), pp. 1578–1585, May 2017. https://doi.org/10.1109/IJCNN.2017.7966039

36. Zhou, B., Khosla, A., Lapedriza, A., Oliva, A., Torralba, A.: Learning deep features for discriminative localization. In: CVPR (2016)

Poster Presentation

Temporal Exceptional Model Mining Using Dynamic Bayesian Networks

Marcos L. P. Bueno[1,2]($\boxtimes$), Arjen Hommersom[3], and Peter J. F. Lucas[4,5]

[1] MCS, Eindhoven University of Technology, Eindhoven, The Netherlands
m.l.de.paula.bueno@tue.nl
[2] iCIS, Radboud University Nijmegen, Nijmegen, The Netherlands
[3] Department of Computer Science, Open Universiteit, Heerlen, The Netherlands
Arjen.Hommersom@ou.nl
[4] Faculty of EEMCS, University of Twente, Enschede, The Netherlands
peter.lucas@utwente.nl
[5] LIACS, Leiden University, Leiden, The Netherlands

Abstract. The discovery of subsets of data that are characterized by models that differ significantly from the entire dataset, is the goal of exceptional model mining. With the increasing availability of temporal data, this task has clear relevance in discovering *deviating temporal subprocesses* that can bring insight into industrial processes, medical treatments, etc. As temporal data is often noisy, high-dimensional and has complex statistical dependencies, discovering such temporal subprocesses is challenging for current exceptional model mining methods. In this paper, we introduce Temporal Exceptional Model Mining to capture multiple and complex relationships among temporal variables of a dataset in a principled way. Our contributions are as follows: (i) we define the new task of temporal exceptional model mining; (ii) we characterize the discovery of exceptional temporal submodels using dynamic Bayesian networks by means of a new distance measure, (iii) we introduce a search procedure for exceptional dynamic Bayesian networks optimized by properties of the proposed distance, and (iv) the practical value of the proposed method is demonstrated based on simulated data and process data of funding applications and by comparisons with other exceptional model mining methods.

Keywords: Machine learning · Graphical models · Bayesian networks · Temporal data · Subgroup discovery · Exceptional model mining

1 Introduction

In many domains such as health care, engineering and workflow processes, there is an increasing availability of temporal data, often mixed with non-temporal ones, such as gender and geographical location. In such cases there may be a need for **discovering subgroups with deviant temporal dynamics** [6,12].

© Springer Nature Switzerland AG 2020
V. Lemaire et al. (Eds.): AALTD 2020, LNAI 12588, pp. 97–112, 2020.
https://doi.org/10.1007/978-3-030-65742-0_7

Examples are male patients for which some symptom takes longer to wane in comparison to female patients, or workflow processes of department A having excessive payment failures in comparison to other departments. This identification is clearly relevant, e.g., to support treatment selection, cost reduction and fraud detection.

As temporal data is often noisy, high-dimensional and has complex statistical dependencies, discovering deviant subprocesses is challenging making many standard statistical and machine learning methods unsuitable. **Exceptional model mining** (EMM) [2,9,10] allows for the discovery of *exceptional* (i.e., deviant) models from temporal data, however restricted to a single temporal observation modeled as a Markov chain (MC) [12]. The MC representation imposes severe limitations for temporal settings, as correlations among multiple observations are invisible as they are collapsed into a single observation. Moreover, scaling to larger problems with MCs is infeasible due to the required number of parameters. On the other hand, temporal submodels with latent variables have been investigated [16], yet interpreting latent states is often not trivial.

One distinguishing feature of EMM is that it supports *interpreting model differences*, explaining *why* an object belongs to a subgroup. The challenge now is: how to represent exceptional temporal subprocesses in EMM with reasonable generality, and yet in an interpretable way? In this paper, we introduce the task of **temporal exceptional model mining** (**TEMM**) for the discovery of exceptional temporal subprocesses. Our definition of TEMM enables the representation of a range of temporal subprocesses. We demonstrate TEMM by means of dynamic Bayesian networks (DBNs) [8] to represent temporal submodels. DBNs are graphical models that fulfill several properties: they can capture arbitrary probability distributions, and are interpretable.

The contributions of this paper are as follows. First, TEMM is presented as a setting for representing exceptional temporal subprocesses in EMM. Then, a distance function that measures the exceptionality of a DBN is introduced. We give a procedure for searching for exceptional DBNs in data that is optimized by exploiting properties of the designed distance. An empirical evaluation demonstrates the proposed method, by a broad comparison with baselines on simulated data and a case based on real workflow process data.

This paper is organized as follows. A running example is described in Sect. 2. In Sect. 3, we define the task of TEMM. In Sect. 4, we introduce a distance measure and a search approach for exceptional DBNs. The experiments based on simulations and real data are discussed in Sects. 5 and 6. In Sect. 7, the related work is reviewed. The conclusions are discussed in Sect. 8.

2   Motivating Example: The Business Process Intelligence Challenge

In the European Union farmers can apply for direct payments, which provide basic income decoupled from production. A *funding application* is described by **Land Area** and **Number Parcels**, is submitted in a **Year** and is handled by a **Department**. The workflow of an application is a set of documents (**Doc Type**), each one having a state (**Subprocess**) that allows for certain actions (**Activity**). For each document, there are one or more subprocesses. This is the basis for the *business process intelligence challenge* (BPIC18) [4].

Typically, the workflow starts with the *payment application* document, with activities such as mail exchange and validation. An application normally requests subsidies for a number of parcels, stored in a *(geo) parcel document*. Checks regarding the validity of parcels are stored in a *department control parcels* document. The stated parcels are also aligned based on a known reference, and this is kept in a *reference alignment*. The result of these and other checks are summarized in the *control summary*. In any document, *editing* and *calculations* are frequent activities. Eventually, a *decision* is made for the case, leading to *payment* activities. Deviations can occur, e.g., a percentage of cases has an *inspection* document with *on-site* or *remote* subprocesses, or the case might also be reopened due to a legal objection. Figure 1 shows this workflow dynamics. Our general goal is to identify the overall dynamics and whether there are subgroups of the data whose dynamics is substantially different from the general one.

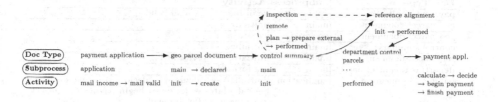

Fig. 1. Typical workflow of the funding example (simplified). Each document occurs with multiple subprocesses and activities. Dashed arrows show process deviations.

3   Temporal Exceptional Model Mining

In this section we describe relevant background notions and define the task of temporal exceptional model mining.

3.1   Temporal Targets

In order to represent subgroups we define descriptor and target variables. The set of descriptor variables is a set $\mathbf{A}$ of random variables $\{A_1, \ldots, A_k\}$, where A_i is a *descriptor variable* and has a domain $\text{dom}(A_i)$. We denote values of the domain by lower-case letters such as $a_i \in \text{dom}(A_i)$. In standard SD, one models next to $\mathbf{A}$ a single variable X called *target variable*, while in EMM a *set of target variables* $\mathbf{X} = \{X_1, \ldots, X_n\}$ is used instead. For example, in EMM for regression [10], the predictor and response variables are the target variables. In TEMM, the target variables $\mathbf{X}$ are the result of a temporal process as defined next.

Definition 1 (Temporal targets). *Let* $\mathbf{X}$ *be a set of random variables. We assume that there is a process that changes* $\mathbf{X}$ *at regular time points, resulting in the variables* $\mathbf{X}^{(0)}, \mathbf{X}^{(1)}, \ldots$ *The variable* $X_i^{(t)}$ *denotes* X_i *at time t, and we denote by* $X_i^{(t_1:t_2)}$ *the variables* X_i *occurring from time* t_1 *up to* t_2. *The variables* $X_i^{(t)}$, *for* $t \geq 0$, *have the same domain. We call each* $\mathbf{X}^{(t)}$ *a temporal target.*

Based on Definition 1, we define the space of variables in TEMM as $\{\mathbf{A}, \mathbf{X}^{(0)}, \mathbf{X}^{(1)}, \ldots\}$. In practice, a data point in TEMM corresponds to configurations of $\mathbf{A}$ and a finite number of temporal targets. Based on this, we consider a multiset D of data points (called dataset in the following), where the ith data point is denoted by $(\mathbf{a}[i], \mathbf{x}[i]^{(0)}, \ldots, \mathbf{x}[i]^{(m_i)})$, in which m_i is its last temporal target.

Example 1. Reconsider the problem of Sect. 2 with descriptors $\mathbf{A} = \{\mathbf{Year}, \mathbf{Department}, \mathbf{Number\ Parcels}, \mathbf{Land\ Area}\}$ and targets $\mathbf{X} = \{\mathbf{Activity}, \mathbf{Doc\ Type}, \mathbf{Subprocess}\}$. Figure 2 shows an example of a data point.

Year = 2016, Department = e7, Number Parcels = 37, Area = 97.85

Doc Type	Subprocess	Activity
payment application	application	mail income → mail valid
geo parcel document	main	initialize
geo parcel document	declared	create
control summary	main	initialize
reference alignment	main	initialize → performed
department control parcels	main	performed
payment application	application	initialize → calculate → decide → revoke decision → calculate → decide → begin payment → insert document → finish payment

(Column on the left labelled: time ↓)

Fig. 2. A data point of the funding process. The temporal targets are {Doc type, Subprocess, Activity}. Arrows indicate transitions between instances of temporal targets. All the activities of a row are associated with the same Doc Type and Subprocess.

3.2 Subgroups

A subgroup can be described by different pattern languages, depending on the data being explored and on the patterns one wishes to discover [5]. Although other languages exist (see, e.g., [2,13]), the attribute-value pattern language is still very relevant in EMM [6,14]. In this work, we use this propositional language, which is defined based on the space of descriptor variables $\mathbf{A}$ as follows.

Definition 2 (Subgroup). *Let* $D = \{d_1, \ldots, d_m\}$ *be a dataset with each records* $d_i = (\mathbf{a}[i], \mathbf{x}[i]^{(0)}, \ldots, \mathbf{x}[i]^{(m_i)})$. *Let* φ *denote an expression of the form* $(A_{p_1} = a_{p_1} \wedge \cdots \wedge A_{p_q} = a_{p_q})$, *where* $\{p_1, \ldots, p_q\} \subseteq \{1, \ldots, k\}$. *The subgroup associated with* φ *is defined as:*

$$G_\varphi = \{d_i \in D \mid (A_{p_1}[i] = a_{p_1} \wedge \ldots \wedge A_{p_q}[i] = a_{p_q})\} \tag{1}$$

We say that the number of descriptors *of* G_φ *is equal to* q.

We refer to a subgroup either by G_φ, by the expression φ that defines it, or simply by G if no confusion arises. For convenience, the domain of a binary descriptor such as A is denoted by $\mathrm{dom}(A) = \{a^-, a^+\}$. For example, an expression $(a_1^+ \wedge a_2^+ \wedge a_3^-)$ represents a subgroup with 3 binary descriptors. In Definition 2, a subgroup is a subset of data points of D selected according to a propositional expression formed by a conjunction of attribute-value pairs. If $q = 1$ we say that the subgroup is *unitary*, otherwise the subgroup is *specialized*.

Definition 3 (Subgroup sequences). *The* subgroup sequences *of a subgroup* G_φ *of D are given by:*

$$S(G_\varphi) = \{\mathbf{x}[i]^{(0:m_i)} \mid d_i \in G_\varphi\} \tag{2}$$

The size *of subgroup G_φ is* $\sum_{d_i \in G_\varphi} (m_i + 1)$ *and is denoted by* $|G_\varphi|$.

In TEMM, given a subgroup G a model shall be fitted on the subgroup's sequences $S(G)$ and is called the *subgroup model*. When we wish to compare subgroups in TEMM, we shall compare the subgroup models associated with these subgroups, hence this comparison is based on the space of temporal targets.

3.3 Problem Statement

In TEMM, we wish to find all the subgroups G whose models have a distribution that differs from the distribution of the subgroup model associated with the rest of the data. Additionally, every subgroup G must have a minimal size, i.e. $|G| \geq \sigma|D|$, where $\sigma \in [0,1]$ is the *minimal size threshold*. One can also specify a preference for more specialized or more general subgroups (see, e.g., [12]).

4 Exceptional Dynamic Bayesian Networks

In this work, dynamic Bayesian networks (DBNs) are studied as model class to represent *temporal* subgroup models. Then, we define a distance notion for DBNs, allowing for the discovery of *exceptional dynamic Bayesian networks*.

4.1 Dynamic Bayesian Networks

Dynamic Bayesian networks extend Bayesian networks (BNs) to model processes with uncertainty [8]: the temporal targets of Definition 1. In order to keep the model compact, a few assumptions are adopted in DBNs. We say that a dynamic system over the temporal targets $\mathbf{X}$ is **Markovian** if $P(\mathbf{X}^{(t+1)} \mid \mathbf{X}^{(0:t)}) = P(\mathbf{X}^{(t+1)} \mid \mathbf{X}^{(t)})$, for all $t \geq 0$. This means that predicting the future state depends only on the current state. Another useful assumption is **time homogeneity**, which holds in a dynamic system if the transitions $P(\mathbf{X}^{(t+1)} \mid \mathbf{X}^{(t)})$ are invariant for every $t \geq 0$.

Definition 4 (Dynamic Bayesian network). *A dynamic Bayesian network M is a Markovian time-homogeneous system $M = (\mathcal{B}_0, \mathcal{B}_\rightarrow)$, where: (i) $\mathcal{B}_0 = (\mathcal{G}_0, P_0)$ is a BN over the variables $\mathbf{X}^{(0)}$ called* **initial network***; (ii) $\mathcal{B}_\rightarrow = (\mathcal{G}_\rightarrow, P_\rightarrow)$ is a BN over the variables $\{\mathbf{X}^{(t+1)}, \mathbf{X}^{(t)}\}$ called* **transition network***. The variables of $\mathbf{X}^{(t)}$ have no parents in the transition network.*

Based on the previous notions, a DBN can be unrolled for any discrete horizon $\{0, \ldots, m\}$ with the following joint distribution:

$$P(\mathbf{X}^{(0:m)}) = \prod_{i=1}^{n} P_0(X_i^{(0)} \mid \pi(X_i^{(0)})) \prod_{t=0}^{m-1} \prod_{i=1}^{n} P_\rightarrow(X_i^{(t+1)} \mid \pi(X_i^{(t+1)})) \quad (3)$$

where $\pi(X_i^{(t)})$ denotes the parents of node $X_i^{(t)}$ in $\mathcal{G}_0$ or $\mathcal{G}_\rightarrow$.

4.2 Distance Function

Definition 5 (Mismatch score). *Let D be a dataset over $\{\mathbf{A}, \mathbf{X}^{(0)}, \mathbf{X}^{(1)}, \ldots\}$ and G, H be two subgroups of D. Further, let us denote by M_G and M_H the dynamic Bayesian networks with maximum score given subgroups G and H respectively. The* **mismatch score** *between M_G and M_H is:*

$$\begin{aligned} mismatch(M_G, M_H) = \; & (score(M_G : G) - score(M_H : G)) \\ & + (score(M_H : H) - score(M_G : H)) \end{aligned} \quad (4)$$

where $score(M : D)$ refers to the score of model M based on subgroup D. Note that given a subgroup G, it holds by definition that $score(M_G : G) \geq score(M_H : G)$ for any model M_H. In practice, it might be difficult to identify the model M_G of subgroup G, as we discuss in Sect. 4.3.

The mismatch score assess the *error* that a model makes when given data different than that which gives the maximum score. Intuitively, if the DBNs of subgroups G and H are similar one would expect a small mismatch, while a high mismatch indicates the models to be highly different.

Proposition 1 (Weak identity of indiscernibles). *Let M_G be the DBN of subgroup G of dataset D. Then it holds that:*

$$mismatch(M_G, M_G) = 0 \quad (5)$$

Note that a mismatch equal to zero does not imply that the subgroups G and H are the same. This is because a dataset D is a multiset, hence G and H might be associated with the same sequences while being two different parts of D.

Proposition 2 (Symmetry). *Given the DBNs M_G and M_H of the subgroups G and H of dataset D, it holds that:*

$$mismatch(M_G, M_H) = mismatch(M_H, M_G) \quad (6)$$

The proofs of Propositions 1 and 2 follow directly from Definition 5.

Proposition 3 (Non-negativity). *Let M_G and M_H be the DBNs of the subgroups G and H of dataset D. Then it holds that:*

$$mismatch(M_G, M_H) \geq 0 \tag{7}$$

Proof. From the assumptions of Definition 5, M_G has the maximum score given G, i.e., $score(M_G : G) \geq score(M_H : G)$ for any model M_H. Analogously, it holds that $score(M_H : H) \geq score(M_G : H)$ for any M_G, which completes the proof.

In the next sections, these properties will appear useful for developing a search strategy for identifying exceptional DBNs.

4.3   Scoring Function

In practice, DBNs can be learned by maximizing a penalized scoring function. In this work, we use the Bayesian information criterion (BIC) [8] as scoring function. The BIC of a model M_G given data G is defined as follows:

$$\text{BIC}(M_G : G) = 2\log \mathcal{L}(M_G : G) - |M_G|\log|G| \tag{8}$$

where $\log \mathcal{L}(M_G : G)$ denotes the log-likelihood of the model M_G, $|M_G|$ the number of parameters of M_G, and $|G|$ is the size of G. We assume that M_G is fitted by maximizing the BIC score on data G. We denote by $\text{BIC}(M_G : H)$, with $H \neq G$, the score of M_G given data H different from data G that was used to fit M_G. The BIC score is the score term in Definition 5.

DBN learning is a hard computational problem. In practice, heuristic search is often used. We refer the reader for further detail on DBN learning [8].

4.4   Exceptional Subgroups

We define next a general notion of exceptional DBNs.

Definition 6 (Exceptional subgroups). *Given a dataset D, we define a relation $ex \subseteq 2^D \times 2^D$, called exceptionality. We say that G is an exceptional subgroup with regard to a subgroup H, denoted by $ex(G, H)$, if the distribution of the DBN M_G is different from the distribution of the DBN M_H.*

It is straightforward to verify that the exceptionality relation just defined is symmetric and anti-reflexive. In EMM, the reference subgroup used for determining the exceptionality of a subgroup is typically the full data D, also referred to as *population* [16]. This means that a subgroup of interest G would be compared with D; however, this comparison is made more convenient by instead comparing G with its complement $\bar{G}$ [5], which results in a comparison involving two disjoint subgroups. This approach will be used in TEMM as well.

4.5   Distribution of False Discoveries

In practice, one way to use Definition 6 for identifying exceptionality is to consider the extent to which subgroup models differ from the population model. In this case, we would like to identify models which are significantly different from the population model. This is because the true distribution of subgroups is unknown, and we therefore need to account for the error in the estimated model.

To determine how exceptional a subgroup G is, a sampling-based approach with the *distribution of false discoveries* (DFD) [7,12] is used. Suppose G has size $|G|$, then random subgroups of size $|G|$ are drawn without replacement from D. The mismatch distance of a random subgroup is computed by fitting a DBN on its data and another DBN on the subgroup's complement data. This procedure approximates the distribution of mismatch distances of subgroups with size $|G|$.

By constructing a distribution of distances of random subgroups, we are able to assess how unusual the mismatch distance of a subgroup G is. In order to do so, we execute a hypothesis testing procedure as follows. By taking large enough number of sampled subgroups, the resulting distribution of random mismatch distances will be approximately Normal (see, e.g., [7,12]). We can then compute a z-score for the mismatch of G, and then a p-value. If the p-value of G is smaller than a significance level α, we conclude that G is an exceptional subgroup.

4.6   Subgroup Search

We introduce a bottom-up search method in Algorithm 1 to identify exceptional subgroups from a dataset D. The central idea of Algorithm 1 is to specialize all exceptional subgroups that have been found so far, until there are no exceptional subgroups to be specialized. Each generated subgroup is predicted as exceptional or non-exceptional using Algorithm 2 (Line 9). The algorithm does not specialize subgroups predicted as non-exceptional. For brevity sake, Line 8 generates several subgroups, one for each value of the new descriptor.

Algorithm 1. Subgroup search

Input: D: a dataset $\{\mathbf{A}, \mathbf{X}^{(0)}, \mathbf{X}^{(1)}, \dots\}$; σ: minimal size threshold; α: significance level.
Output: E: set of subgroups predicted as exceptional.

1: $E \leftarrow \emptyset$
2: $F \leftarrow \emptyset$ // Exceptional subgroups to further expand
3: $C \leftarrow \emptyset$ // Current subgroup
4: cand_descs $\leftarrow \{A_1, \dots, A_k\}$
5: **do**
6:     $E' \leftarrow \emptyset$
7:     **for all** $A_i \in$ get_cand_descriptors(c) **do**
8:         $G \leftarrow C \cup \{A_i = a_i\}$, for each $a_i \in \text{dom}(A_i)$ // Specialize current subgroup C
9:         **if** check_size(G, D, σ) and exceptionality_test(G, D, α) **then**
10:            $E' \leftarrow E' \cup \{G\}$
        // Add new exceptionals and select new one for expansion
11:     $E \leftarrow E \cup E'$
12:     $F \leftarrow F \cup E'$
13:     $C \leftarrow$ select_random(F)
14:     $F \leftarrow F - \{C\}$
15: **while** $F \neq \emptyset$
16: **return** E

4.7 Exceptionality Test

Algorithm 2 predicts the exceptionality of a subgroup using the statistical test of Sect. 4.5. The test assesses how unusual the mismatch of a subgroup is compared to the distribution of mismatch distances of random subgroups, by constructing a DFD. We sample 100 subgroups in our experiments to build each DFD.

Computing a DFD from the scratch is costly due to multiple DBN learning calls. However, we can avoid this by noting that the DFD is a function of the subgroup size, hence when asking for the DFD of a subgroup G we can *reuse* the previously computed DFD of a subgroup H if $|G| = |H|$, which enables substantial computation savings. Moreover, by Proposition 2 the mismatch distance is symmetric, hence when we look up for a DFD in our table of stored DFDs, we can look up for DFDs associated with size $|G|$ *and* to DFDs associated with size $|D| - |G|$. This yields additional computation savings.

Algorithm 2. Exceptionality test

Input: G: a subgroup; D: a dataset $\{A, X^{(0)}, X^{(1)}, \ldots\}$; α: significance level.
Output: the exceptionality prediction of G.

1: $M_G \leftarrow$ learn_dbn($S(G)$)
2: $M_{\bar{G}} \leftarrow$ learn_dbn($S(\bar{G})$)
3: $d \leftarrow$ mismatch($M_G, M_{\bar{G}}$)
   // Distribution of false discoveries
4: **if** dfd_exists($|G|$) **then** // By Proposition 2, also search for a DFD with size $|D| - |G|$
5:     $d_s \leftarrow$ get_stored_mismatch_distances($|G|$) // Reuse DFD
6: **else** // Reuse not possible: compute DFD from scratch
7:     Sample subgroups from D with size $|G|$ and make $d_s \leftarrow \emptyset$
8:     **for all** sampled subgroup H **do**
9:         $M_H \leftarrow$ learn_dbn($S(H)$)
10:        $M_{\bar{H}} \leftarrow$ learn_dbn($S(\bar{H})$)
11:        $d_H \leftarrow$ mismatch($M_H, M_{\bar{H}}$)
12:        $d_s \leftarrow d_s \cup \{d_H\}$
13:    store_mismatch_distances($|G|, d_s$)
14: Calculate the mean $\bar{x}$ and standard deviation s from the set of distances d_s
15: $z \leftarrow \dfrac{d - \bar{x}}{s}$ // z-score of the subgroup
16: Calculate the p-value corresponding to the z-score.
17: **if** p-value $< \alpha$ **then**
18:    **return** *true* // Subgroup predicted as exceptional
19: **return** *false* // Subgroup predicted as non-exceptional

---

5 Experiments with Simulated Data

5.1 Data Generating Procedure

We consider two *simulation scenarios* for assessing the method[1]. First, the number of temporal targets n in $X = \{X_1, \ldots, X_n\}$, with X_i binary, is set to $n = 10$ inspired by previous research [12] which used Markov chains with 1,024 states. Second, we consider 100 times more states for a broader evaluation, requiring

---

[1] Source code and datasets available at: https://github.com/marcoslbueno/temm.

$n = \log_2 100 \cdot 1024 \simeq 17$ temporal targets. For each scenario, two ground truth DBNs on $\mathbf{X}$ were built, with model structure generated by uniformly sampling directed acyclic graphs and node parameters sampled from Beta distributions. Data sequences were sampled from the DBNs, with duration of 10 time points. The same amount of data was sampled from each DBN.

Next, we include the descriptor variable A_1 such that $A_1 = a_1^-$ for all the sequences from one DBN, and $A_1 = a_1^+$ for all the sequences of the other DBN. We also added 5 binary descriptors $R_1, \ldots, R_5$ to act as noisy variables, such that the value of R_i on each sequence is assigned uniformly at random. Based on this procedure, simulated data for a scenario consists of data points over $\{A_1, R_1, \ldots, R_5, \mathbf{X}^{(0)}, \ldots, \mathbf{X}^{(9)}\}$, where $m = 9$ (the last time point) and the cardinality of $\mathbf{X}$ is n.

5.2   Evaluation

The ultimate goal of TEMM is to recover the exceptional subgroups. For evaluation purposes, we see this as a classification problem on the *space of descriptors*, such that each subgroup is either a *positive* or a *negative* instance. We assigned *ground truth labels* to unitary subgroups as follows:

- *Positive instances*: subgroups (a_1^+) and (a_1^-), as the sequences of each come from different DBNs, making these subgroups exceptional by definition.
- *Negative instances*: subgroups described by R_i, such as (r_1^+) and (r_1^-) as they contain sequences from both DBNs selected at random.

The predicted labels of unitary subgroups by Algorithm 1 are used to evaluate the proposed method. The AUROC (area under the ROC curve) was computed, allowing us to measure how well we can identify exceptional subgroups. We also evaluated the specialized subgroups that Algorithm 1 generates if exceptional unitary subgroups are found. Analogously, positive instances are specialized subgroups that include A_1, and negative instances are all the other specialized subgroups. We evaluate unitary and specialized subgroups separately as the number of specialized ones is typically much larger.

Baseline. Markov chains were used as baseline for representing the temporal targets instead of DBNs. For both MC and DBNs, we applied the mismatch score from Definition 5 to identify subgroups. To avoid zero probabilities, Laplace smoothing with smoothing parameter $\lambda = 1$ is used in both MC and DBN parameter estimation. The whole simulation process was executed 10 times for better assessment, each time with different ground truth models.

5.3   Results

Figure 3a shows the results based on simulated data for unitary subgroups. Note that the X axis shows the number of sequences in each ground-truth subgroup, hence the total dataset size is twice that amount. The results suggest that the

DBN and the MC representation achieved good results with datasets of $n = 10$ target variables (or 1,024 MC states). However, substantial differences arose with $n = 17$ variables (or 131,072 MC states), a situation where DBNs were able to provide optimal AUC values even with the minimal amount of data, as opposed to MCs. In this case, MCs had to count on substantially larger amounts of data in order to provide comparable AUC values to those of DBNs. The threshold $\alpha = 0.05$ was used in Algorithm 2.

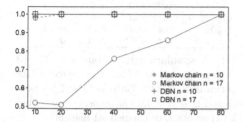
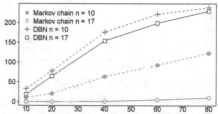

(a) Number of sequences (X axis) and mean AUROC (Y axis) on *unitary* subgroups.

(b) Mean number of exceptional *specialized* subgroups correctly predicted (Y).

Fig. 3. Results of Markov Chains and DBNs on simulated data (10 simulations).

Figure 3b shows the mean number of specialized subgroups which include A_1 and were labeled as exceptional. As the amount of data increases, the results show that more subgroups were produced by both the MC and DBN representations. However, it is clear that DBNs were able to capture substantially more specialized exceptional subgroups.

Figure 4b shows a fragment of subgroups from a simulation iteration using DBNs, together with their mismatch distances. This shows that the method is robust at identifying exceptional subgroups even when most of other subgroups are noisy subgroups. Moreover, the mismatch distances of exceptional subgroups are usually very different from those of non-exceptional subgroups.

5.4   Impact of (dis)similar Models on Prediction

Now we consider simulations where we control the similarity of the ground truth models. To this end, the second ground truth DBN was defined by copying the structure and parameters of the first DBN. Then, for each variable X_i in the second DBN let $p \leftarrow P(X_i^{(0)} = x_i^- \mid \pi(x_i^{(0)}))$ and $p' \leftarrow P(X_i^{(0)} = x_i^+ \mid \pi(x_i^{(0)}))$. Then, these parameters are changed by picking at random a real number called *change* from the interval $[0, \min(\delta, 1 - p)]$, with uniform probability, where $\delta \in [0, 1]$ is the *maximal change threshold*. Next, we set $p \leftarrow p + change$ and $p' \leftarrow p' - change$. The lower the threshold δ, the more similar the DBNs are.

Except for the way ground truth DBNs are generated, we follow the data generating procedure of Sect. 5.1 and restrict ourselves to learning DBNs and use $n = 17$ temporal targets. Figure 4a shows the AUROC of simulations based on different δ values. The results suggest that extreme cases (low δ, little data) are challenging for the proposed method. In the remainder cases, the method achieved good to optimal results, which suggests that the method is robust at detecting exceptional behavior.

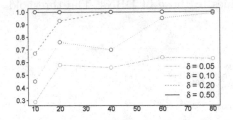

Subgroup	Size	z-score	p-value	Labels (P&T)	
(a_1^+)	0.50	195.8	$\simeq 0$	1	1
(a_1^-, r_2^-)	0.27	49.4	$\simeq 0$	1	1
(a_1^+, r_1^+, r_2^+)	0.11	15.1	$\simeq 0$	1	1
(r_2^-)	0.49	-1.2	0.22	0	0
(r_3^-)	0.49	0.5	0.64	0	0

(a) X axis: Number of sequences in dataset; Y axis: AUROC for different values of δ (maximal change threshold).

(b) A simulation iteration ($n = 17$, 80 data sequences). **Size** = subgroup size normalized by $|D|$, **Labels (P&T)** = predicted and true labels respectively. Label 1 (0) = positive (negative) instance.

Fig. 4. Results of DBNs on simulated data with varying similarity of ground truth.

6   Data of Funding Applications

In order to evaluate the proposed TEMM method, we consider data from the *business process intelligence challenge* (BPIC18) [4], already briefly described in Sect. 2. The BPIC18 dataset contains event log data of applications submitted to the European Union for direct payments for German farmers in 2015–2017. *The goal of applying TEMM to the BPIC18 data is to identify subgroups whose dynamics of events is exceptional.*

6.1   Data

Each application in the BPIC18 data is associated with descriptor variables (domain size) as follows: **Land Area** (437), **Department** (4), **Number of Parcels** (74), **Redistribution** (2), **Year of Submission** (3), **Success** (2), **Small Farmer** (2), and **Young Farmer** (2). Applications are also associated with *events* related to workflow activities, where an event is described by the multinomial variables (domain size): **Doc Type** (8), **Subprocess** (8) and **Activity** (33). From the original set of 41 activities, we filtered out some repetitive and generic activities, such as *editing* and *save*.

Each application is associated with one or more events, which are the temporal targets of the data. Hence, the ith data point of this dataset has the form {Land Area, ..., Young Farmer, Activity$^{(0:m_i)}$, ..., Subprocess$^{(0:m_i)}$}. The BPIC18 dataset has 4,800 applications randomly selected from the original dataset, with an equal number of applications per year. There are 145,980 events in total (mean [StDv] length of each application: 30.4 [8.4] events). Again, Laplace smoothing with $\lambda = 1$ was used in model learning.

6.2 Discovered Subgroups

Table 1a shows an excerpt of the exceptional subgroups discovered from the BPIC18 data based on a minimal size $\sigma = 0.05$. The results show that the most exceptional subgroups are unitary and described by a particular year, be it 2015, 2016 or 2017. This suggests that significant changes took place in application dynamics across years manifested in the sequential behavior of the target variables. This could be explained, e.g., by changes in the business process and funding policies. Each department also has its own dynamics, as all unitary subgroups with this descriptor were exceptional. However, their exceptionality was not as strong as that of year subgroups.

Table 1. Results on the BPIC18 dataset, where 38 exeptional subgroups were discovered. For better visualization, only the 5 most exceptional specialized subgroups are shown. All p-values < 0.001, except (Number Parcels=1).

Exceptional subgroups	Size	z-score
Year=2015	0.37	2461.47
Year=2016	0.33	1327.07
Year=2017	0.30	2411.69
Department=4e	0.32	33.28
Department=e7	0.28	35.03
Department=6b	0.25	24.29
Department=d4	0.16	28.00
Number Parcels=2	0.06	12.15
Number Parcels=3	0.06	25.10
Number Parcels=1	0.05	2.15
Year=2015 ∧ Young Farmer=False	0.34	2107.47
Year=2017 ∧ Young Farmer=False	0.27	1844.72
Year=2016 ∧Young Farmer=False	0.30	1144.32
Department=4e ∧Year=2015	0.11	730.81
Department=e7 ∧ Year=2015	0.11	647.71

(a) **Size** = subgroup size normalized by $|D|$.

Doc Type	2015	2016	2017
payment application	16	20.8	12.1
entitlement application	10.5	0.3	0.1
parcel document	2.6	0	0
control summary	1	1	1
reference alignment	2.2	2.1	2
department control parcels	1	1	0
inspection	0.6	1	0.8
geo parcel document	0	3.8	11.3

(b) Average number of document types per application in each year.

6.3 Comparison to Previous Analyses

While the ground truth exceptional subgroups are not available for the BPIC dataset, there is evidence that the subgroups described by *year* as shown in Table 1a are exceptional. First, the BPIC18 data provider [4] claims that the underlying process changed between years due to changes implemented in the structure of the application procedure. This is in line with previous research [15] on this dataset, where concept drifts were identified precisely between each year of the data. Other research [19] has analyzed how the workflow of applications submitted in different years has changed, also suggesting that differences exist in these workflow structures. Based on these previous analyzes, we conclude the proposed method is able to detect true exceptional subgroups.

Differently than the other analyses from the literature on the BPIC18 data, the method proposed in this paper can be seen as a principled one due to its statistical foundations.

6.4 Subgroup Differences

Based on subgroup's data, Table 1b shows the frequency of each Doc Type value for the most exceptional subgroups. One strong difference is that the *geo parcel document* vanished in applications from 2015, while it was increasingly used in applications from 2016 and 2017. On the other hand, the *parcel document* was adopted only in 2015, and the *document control parcels* vanished in 2017. All these changes are expected due to known changes in the funding process [4].

Table 1b also reveals a remarkable reduction in the frequency of *entitlement application* over the years. This could reflect that subprocesses such as *objection* and *change* of entitlement application are moved to application payment, as the latter is the only other type of document which has such subprocesses. Other changes include more *inspections* in 2016 and 2017, which might indicate changes in funding policies as only a small percentage of cases are to be inspected.

7 Related Work

As a generalization of SD, exceptional model mining [6] is an active area of research and has been applied to different target variable representations. Earlier research includes the discovery of exceptional linear regression models [10] and the discovery of subgroups with Bayesian networks that have significant structural differences [5]. A more specialized usage of EMM is tailored at sequential problems, yet over a single target, where discrete Markov chains with significantly different transition patterns have been investigated [12].

The aforementioned EMM research can be seen as *parameter-based approaches*, because subgroups are characterized based on the unusualness of model parameters, such as regression slope and network structure. On the other hand, model-based subgroup discovery [16] is an *evaluation-driven approach* that compares the distribution of subgroups by means of proper scoring rules. The latter is related to data mining research where the minimum description length (MDL) was applied to identify differences between databases [18]. In this paper we consider more general model selection criteria, where MDL is a special case.

Some body of research has dealt with *subgroup search*, whose aims include making the search more efficient and reducing the number of redundant subgroups. Research has been done on providing bounds for some interestingness scores in the context of numerical targets that can be used for search pruning [11]. Subgroup search has also been formulated in terms of game theory [3], which allows for guiding the search toward the interestingness of subgroups while improving the lack of diversity that search might face.

Other extensions to SD and EMM operate on data other than the common attribute-value data. The approach in [13] is tailored for relational data and can extract very general structured patterns of subgroups. More recently, exceptional graph mining [2] has been proposed to allow for the discovery of graph neighborhoods that are similar internally but exceptional to the general attributed graph (i.e. graphs with non-trivial vertices such as a list of attribute-value pairs). Recently, EMM has been applied to finding subsets of data related to exceptional convolutional layers in convolutional neural networks [17], which might help the interpretation of such models.

The proposed mismatch score can be seen as a *data-based* score, as it is computed based on goodness-of-fit scores (the BIC score). By opposition, previous research [12] for discovering exceptional MCs used a measure based on statistical distance between transition distributions. While structure learning is not required for MC learning, the number of parameters in DBNs is typically substantially lower due to its factorized representation. As experiments have shown, this parameter issue makes the MC representation to scale poorly, particularly when the number of temporal targets n is larger and there is a less data for model learning. Furthermore, the DBN-based search made substantially less mistakes in the simulations, which makes this representation suitable for TEMM.

One task that has some resemblance to TEMM is sequential pattern mining [1]. However, the mined rules might not correspond to actual subgroups or even actual processes from the dataset, as opposed to TEMM and subgroup discovery. This makes it not possible to directly compare the results of these approaches.

8 Conclusions

In this paper, we proposed temporal exceptional model mining to enable the representation of temporal observations in EMM in a principled way. For capturing the temporal dependencies in TEMM, dynamic Bayesian networks were used, which allows for an intuitive and interpretable model class for TEMM.

The proposed method was empirically evaluated on simulated data and process data based on funding applications, showing that the identifiability of the method in different scenarios is robust. Our method was able to discover exceptional subgroups from the funding data in accordance to previous research, as well other, yet less exceptional subgroups. Furthermore, our approach solved this practical problem in a more principled manner.

As future work, we would like to explain in more detail why models are considered as exceptional. This could involve looking at relevant structural or numerical parameters of the DBNs. We wish to quantify the savings of the optimizations employed during search to reduce the computation of distribution of false discoveries. Finally, we would like to investigate if further improvements to the search algorithm are possible based on properties of the mismatch distance.

References

1. van der Aalst, W.: Process Mining: Discovery, Conformance and Enhancement of Business Processes. Springer, Heidelberg (2011). https://doi.org/10.1007/978-3-642-19345-3
2. Bendimerad, A., Plantevit, M., Robardet, C.: Mining exceptional closed patterns in attributed graphs. Knowl. Inf. Syst. **56**(1), 1–25 (2018). https://doi.org/10.1007/s10115-017-1109-2

3. Bosc, G., Boulicaut, J.F., Raïssi, C., Kaytoue, M.: Anytime discovery of a diverse set of patterns with Monte Carlo tree search. Data Min. Knowl. Discov. **32**(3), 604–650 (2018). https://doi.org/10.1007/s10618-017-0547-5
4. van Dongen, B., Borchert, F.: BPI Challenge 2018 (2018). https://data.4tu.nl/repository/uuid:3301445f-95e8-4ff0-98a4-901f1f204972
5. Duivesteijn, W., Knobbe, A., Feelders, A., van Leeuwen, M.: Subgroup discovery meets bayesian networks - an exceptional model mining approach. In: 2010 IEEE International Conference on Data Mining. pp. 158–167, December 2010
6. Duivesteijn, W., Feelders, A.J., Knobbe, A.: Exceptional model mining. Data Min. Knowl. Discov. **30**(1), 47–98 (2016)
7. Duivesteijn, W., Knobbe, A.: Exploiting false discoveries - statistical validation of patterns and quality measures in subgroup discovery. In: Proceedings of the IEEE 11th International Conference on Data Mining, ICDM 2011, pp. 151–160 (2011)
8. Friedman, N., Murphy, K., Russell, S.: Learning the structure of dynamic probabilistic networks. In: Proceedings of the 14th Conference on Uncertainty in Artificial Intelligence, UAI 1998 (1998)
9. Herrera, F., Carmona, C.J., González, P., del Jesus, M.J.: An overview on subgroup discovery: foundations and applications. Knowl. Inf. Syst. **29**(3), 495–525 (2011). https://doi.org/10.1007/s10115-010-0356-2
10. Leman, Dennis., Feelders, Ad, Knobbe, Arno: Exceptional model mining. In: Daelemans, Walter, Goethals, Bart, Morik, Katharina (eds.) ECML PKDD 2008, Part II. LNCS (LNAI), vol. 5212, pp. 1–16. Springer, Heidelberg (2008). https://doi.org/10.1007/978-3-540-87481-2_1
11. Lemmerich, F., Atzmueller, M., Puppe, F.: Fast exhaustive subgroup discovery with numerical target concepts. Data Min. Knowl. Disc. **30**(3), 711–762 (2016)
12. Lemmerich, F., et al.: Mining subgroups with exceptional transition behavior. In: Proceedings of the 22nd ACM SIGKDD International Conference on Knowledge Discovery and Data Mining, KDD 2016, pp. 965–974 (2016)
13. Lemmerich, F., et al.: Mining subgroups with exceptional transition behavior. In: Proceedings of the 22nd ACM SIGKDD International Conference on Knowledge Discovery and Data Mining, KDD 2016, pp. 965–974 (2016)
14. Novak, P.K., Lavrač, N., Webb, G.I.: Supervised descriptive rule discovery: A unifying survey of contrast set, emerging pattern and subgroup mining. J. Mach. Learn. Res. **10**, 377–403 (2009)
15. Pauwels, S., Calders, T.: An anomaly detection technique for business processes based on extended dynamic Bayesian networks. In: The 34th ACM/SIGAPP Symposium on Applied Computing (SAC 2019), pp. 1–8. Limassol, Cyprus (2019)
16. Song, H.: Model-Based Subgroup Discovery. Ph.D. thesis, University of Bristol (11 2017)
17. van Strien, B.: Exceptional Model Mining of Convolutional Neural Networks. M.Sc. thesis, Eindhoven University of Technology (2019)
18. Vreeken, J., Van Leeuwen, M., Siebes, A.: Characterising the difference. In: Proceedings of the 13th ACM SIGKDD International Conference on Knowledge Discovery and Data Mining, pp. 765–774 (2007)
19. Wangikar, L., Dhuwalia, S., Yadav, A., Dikshit, B., Yadav, D.: Faster Payments to Farmers: Analysis of the Direct Payments Process of EU's Agricultural Guarantee Fund - Business Process Intelligence Challenge 2018 (2018)

"J'veux du Soleil"
Towards a Decade of Solar Irradiation Data (La Réunion Island, SW Indian Ocean)

Mathieu Delsaut[1], Patrick Jeanty[1], Béatrice Morel[1], and Dominique Gay[2]($\boxtimes$)

[1] LE2P-EA4079. Université de La Réunion, Réunion, France
{mathieu.delsaut,patrick.jeanty,beatrice.morel}@univ-reunion.fr
[2] LIM-EA2525. Université de La Réunion, Réunion, France
dominique.gay@univ-reunion.fr

Abstract. This paper aims at presenting years of solar irradiation data together with meteorological data acquisition localized in the French region of La Réunion Island (SW Indian Ocean). The publicly available data take the form of multivariate time series data with one-minute sampling rate over eight years – with still ongoing acquisition. We also present typical analytics tasks that are related to solar energy application domain as well as general time series analytics tasks that are suitable for these data. Thus, we aim at drawing the attention of the time series data mining community to these valuable data.

Keywords: Solar irradiation data · Multivariate time series data · Open data

Preamble. This paper is a *resource track paper*. Its aim is mainly to describe an innovative data set to *(i)* support research on the solar irradiation topic; *(ii)* to potentially suggest novel evaluation tasks; *(iii)* to encourage novel methods and/or algorithms. The concerned data set is already available under reasonably liberal terms and we hope sufficiently well-documented. We also suggest some open research and valuable applications.

1 Introduction

The European Union long-term climate strategy aims to be climate-neutral by 2050, i.e., an economy with net-zero greenhouse gas emissions [3]. As one of the most remote regions of EU, La Réunion island, a French overseas department, has implemented a EU-consistent multiannual energy program which main goal is *electric energy self-sufficiency* in the horizon 2030 [6]. Indeed, as an island (SW Indian Ocean), its isolated position prevents from being interconnected with the metropolitan power grid – thus leading to high dependency from fossil fuels for electricity production.

However, as a tropical island, La Réunion presents a high potential of renewable energies: besides biomass and wind exploitation, solar resource has attracted much

© Springer Nature Switzerland AG 2020
V. Lemaire et al. (Eds.): AALTD 2020, LNAI 12588, pp. 113–121, 2020.
https://doi.org/10.1007/978-3-030-65742-0_8

attention in the past decade. In order to estimate the solar energy potential of the island, a local research team has led a solar resource research programme for the use of solar resource as a stable source of energy and to ensure its management in a reliable and efficient way for its integration into an electrical power grid [4].

In this paper, we present the data acquired during the last decade through the research programme. Sensors used for solar irradiation and meteorological parameters measurement are presented in Sect. 2. Section 3 is dedicated to the full description of the publicly available data. We discuss the main domain applications using the available data in Sect. 4 before concluding.

2   Data Acquisition

In order to obtain a representative view of the solar energy potential, several measurement stations has been spread out over the island. More precisely, six stations have been installed, mainly on EDF[1] power plants sites located along the leeward coast of the island which also concentrates most of the inhabitants (see Fig. 1):

1. Moufia (*reference station* of the Université de La Réunion, Saint-Denis)
2. Bois de Nèfles (EDF site, Saint-Denis)
3. Saint-André (EDF site, Saint-André)
4. Port-Est (EDF site, Le Port)
5. Saint-Leu (EDF site, Saint-Leu)
6. Saint-Pierre (EDF site, Saint-Pierre)

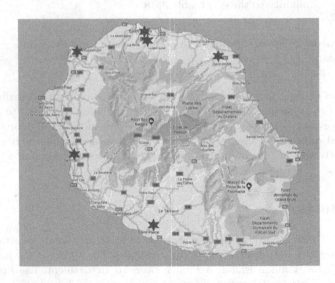

Fig. 1. Localisation of the 6 measurement stations over La Réunion.

---

[1] Electricité De France.

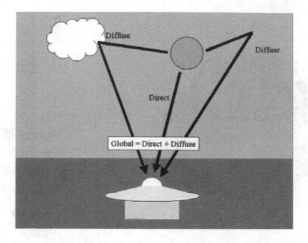

Fig. 2. Global, direct and diffuse components of solar irradiance.

2.1 Measuring Irradiation and Meteorological Parameters

Each station has roughly the same set-up (hardware and software) that can be split into three main parts: the measuring tools, the control system and the supporting structure. Here, we focus on the measuring tools while more technical setup are available in Appendix A.

Measuring tools - Ground-based measurements come from two sensors:

- 1 **pyranometer** SPN1 (manufacturer: Delta-T Devices) which simultaneously measures:
    - the **GHI** (Global Horizontal Irradiance in W/m^2)
    - and the **DHI** (Diffuse Horizontal Irradiance in W/m^2).

  These two components of solar irradiation are illustrated in Fig. 2.
  Finally, the **BHI** (Beam Horizontal Irradiance in W/m^2) may be easily obtained by difference of the previous two.

$$GHI = BHI + DHI$$

One can also compute the **DNI** (Direct Normal Irradiance in W/m^2) by introducing the zenith angle Θ:

$$DNI = \frac{BHI}{\cos \Theta}$$

- 1 **weather transmitter** WXT520 (manufacturer: Vaisala) which measures five meteorological parameters:
    - air temperature (°C),
    - atmospheric pressure (Pa),
    - relative humidity (%),
    - wind speed (m/s)
    - and wind direction (°)

These two measurement tools are illustrated in Fig. 3.

Fig. 3. Main measurement tools: (left) SPN1 pyranometer and (right) WXT520 weather transmitter.

3   Available Data

The six stations have been installed in 2012. Thus, eight years of historical data are now available. Considering the seven parameters, (GHI and DHI, plus 5 meterological parameters), the data takes the form of multivariate time series (MTS). As the sampling rate is one minute, the MTS data contains more than 150 millions data points. A short extract in csv format is shown in Fig. 4.

Timestamp	FD_Avg	FG_Avg	Patm_Avg	RH_Avg	Text_Avg	WD_MeanUnitVector	WS_Mean
...							
01/05/14 10:20	72.32	687.6	980.2	58.26666	26.78333	308.8434	0.5833333
01/05/14 10:21	71.7	688.1	980.2	56.83333	26.9	330.7189	0.9333333
01/05/14 10:22	71.36	673.7	980.2	56.76666	26.91666	310.8201	0.6833333
01/05/14 10:23	72.95	666.3	980.1333	57.01666	27.0	345.3426	2.216666
01/05/14 10:24	75.5	672.0	980.0999	56.66667	26.93333	355.3273	2.233333
01/05/14 10:25	78.67	680.8	980.0999	57.56667	26.76666	340.6307	1.216667
01/05/14 10:26	76.8	683.2	980.0999	57.96666	26.7	318.039	0.6166667
01/05/14 10:27	75.27	689.0	980.0999	57.75	26.71667	23.57079	1.083333
01/05/14 10:28	74.08	692.7	980.0166	58.25	26.8	23.79196	0.65
01/05/14 10:29	72.89	710.5	980.0333	58.4	26.8	24.55521	0.9
01/05/14 10:30	72.1	709.6	980.0	57.46666	26.8	341.5136	1.416667
...							

Fig. 4. Multivariate data sample in csv format for a short 10-min. period on may 1st, 2014. Timestamp (date and time), FD_Avg (diffuse), FG_Avg (global), Patm_Avg (atmospheric pressure), RH_Avg (relative humidity), Text_Avg (Temperature), WD_MeanUnitVector (wind direction) and WS_Mean (wind speed).

Typical Irradiation Data Shapes - Given the tropical climate of the island, at least four typical days might be observed when regarding GHI and DHI parameters. We illustrate these cases in Fig. 5: (a) a sunny day is identified by a hill-form GHI curve (with the maximum around noon) and a high difference with DHI almost flat curve; (b) a cloudy day when GHI is confounded with DHI; (c) intermittency (e.g., frequent cloud pass) is characterized by a high variability in GHI; and (d) a sequence of the three previous phenomenons.

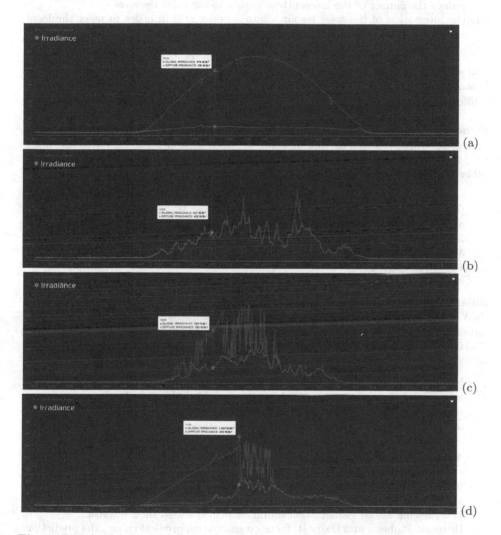

Fig. 5. GHI (red) and DHI (orange) evolution during four typical days : (a) sunny, (b) cloudy, (c) intermittent then cloudy, (d) sequence of sunny, intermittency, cloudy periods. (Color figure online)

4   Valuable Applications

Solar Energy Domain - As pointed out in [4], using solar resource as an electrical and stable source of energy is not an easy task. It raises three scientific and technical issues that are still open:

1. the prediction of solar irradiation despite its variability
2. the management of the solar photovoltaic production through storage systems to reduce the impact of the intermittent nature of the solar resource
3. the integration of the solar resource into a power grid in order to meet the local energy needs and to cope with the load fluctuations

The first problem, the prediction of solar irradiation, is particularly important as it is a precondition for the success of the two others. From a data science point of view, it may directly be formalized as a forecasting problem when considering either short, mid or long term prediction (e.g., 1 h, 6 h or a day ahead forecasting).

General Machine Learning/Data Mining - Besides daily clustering or short, mid, long term forecasting of solar irradiation, which correspond to core ML/DM tasks for MTS, one can benefit from this large-scale MTS data source for the evaluation of other *classical* ML/DM tasks like, e.g.:

- correlation analysis between irradiation parameters and meteorological parameters (more generally, between the various dimensions of MTS)
- outlier detection/extreme values analysis for MTS
- MTS missing values imputation,
- MTS data compression
- similarity-based query optimisation
- advanced visualisation techniques for large-scale MTS data

Solar Irradiation Data: a Challenging Data Set - A two-year piece of these data (2014–2015) has been suggested as an open challenge data set for the annual data mining challenge of the 2018 French data mining and management conference (EGC 2018) [2]. For this open challenge, two selected and award-winning papers have addressed the irradiation data clustering and prediction problems:

- Per, Dalleau and Smail-Tabbone [5] has explored the challenge data in multiple ways: Prior exploratory data analyses have enabled the statistical comparison of characteristics of cities with respect to the measured weather variables (diffuse and overall solar fluxes, atmospheric pressure, moisture, temperature, wind speed and direction). Data was preprocessed and univariate time-series and multivariate time-series aggregated over hours or days were analyzed in order to build simple and effective prediction models. A classical clustering approach was performed. Groups of days sharing weather parameters in common were found by two biclustering algorithms. The characterisation of found biclusters and their succession displayed in a calendar-based visualization tool have helped assess their interest.
- Bruneau, Pinheiro and Didry [1] focus on short-term prediction, i.e., the prediction of solar irradiance one hour ahead. The authors have tested the value of using recently observed data as input for prediction models, as well as the performance of models across sites. After a data cleaning and normalization pre-processing step, they combine a variable selection step based on AutoRegressive Integrated Moving Average (ARIMA) models, to end up with general purpose regression techniques such as neural networks and regression trees.

5   Conclusion

In this data paper, we have presented eight years of solar irradiation data together with meteorological data. The raw data takes the form of multivariate times series with one-minute sampling rate. While the data acquisition is still active, we think the already available data is a valuable support for advanced analytical tasks such as daily solar irradiance clustering or forecasting, as well as other general analytical tasks. Thus, we hope to see many papers using these data in the future.

Acknowledgments. Publicly available data come from recent successive research projects, RCI-GS and GeoSun, with financial support from Europe, Regional Reunion Island Council and the French government through the ERDF (European Regional Development Fund).

Permanent link to La Réunion Island solar irradiation Data -
https://doi.org/10.5281/zenodo.3898530

A   Technical Setup

Data logging - The control system contains the following equipment (see Fig. 6):

CR 1000

NL 115

230 VCA – 15 VDC converter

power regulator

rack for battery (when necessary)

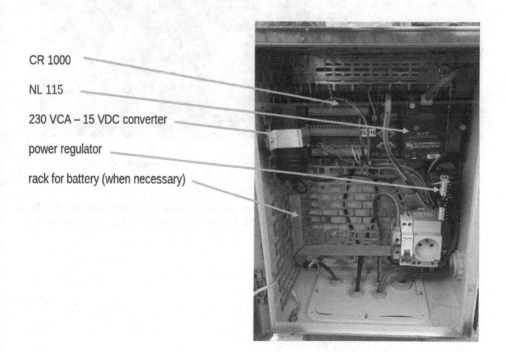

Fig. 6. Equipment of a control system unit.

- 1 waterproof IP64 enclosure (manufacturer: Legrand), size 400 mm × 300 mm × 250 mm, in fiberglass or coated metal.
- 1 datalogger CR1000 (manufacturer: Campbell Scientific), a programmable device that handles sensors measurements, drives communication and stores data and programs.
- 1 Ethernet interface NL 115 (manufacturer: Campbell Scientific) with memory module.
- 1 CompactFlash memory card 500 MB

The station may be powered through 230 VCA grid or by renewable energy using solar panel and battery. Communication between station and server may be ensured by a GPRS link (in that case, a modem with a SIM card needs to be added in the local cabinet in connection with the CR1000 logger).

Supporting Structure - Ground-based stations network hosts two types of stations depending of environmental conditions. In EDF power plant sites, a 10-m high foldable mast (see Fig. 7) is used as well as a metallic enclosure.

Fig. 7. 10-m high foldable mast supporting a measurement station.

In other places, a compact and easily transportable station is used (see Fig. 8). In any case, the whole station is designed and fabricated to bear cyclones winds up to 250 km/h.

Fig. 8. A compact and easily transportable station setup.

References

1. Bruneau, P., Pinheiro, P., Didry, Y.: Prédiction du rayonnement solaire par apprentissage automatique. In: Lebbah, M., Largeron, C., Azzag, H. (eds.) Extraction et Gestion des Connaissances, EGC 2018, Paris, France, January 23–26, 2018. RNTI, vol. E-34, pp. 439–450. Éditions RNTI (2018)
2. Défi EGC 2018: Un défi sous le soleil de l'île de la Réunion (2018). https://www.egc.asso.fr/manifestations/defi-egc/defi-egc-2018.html (in French)
3. European Union - Climate strategies and targets - 2050 long-term strategy (2020). https://ec.europa.eu/clima/policies/strategies/2050_en
4. Goujon Person, M., Delage, O., Bessafi, M., Chabriat, J.P., Jeanty, P.: Solar energy research and development program on the exploitation of the solar resource on the Réunion island and its integration into an electrical power grid. In: Third Southern African Solar Energy Conference (SASEC 2015), Skukuza, South Africa, May 2015
5. Per, Y., Dalleau, K., Smaïl-Tabbone, M.: Exploration et analyses multi-objectifs de séries temporelles de données météorologiques. In: Lebbah, M., Largeron, C., Azzag, H. (eds.) Extraction et Gestion des Connaissances, EGC 2018, Paris, France, January 23–26, 2018. RNTI, vol. E-34, pp. 427–438. Éditions RNTI (2018)
6. Région Réunion - multiannual energy program - programmation pluriannuelle de l'énergie (ppe) (2020). http://www.reunion.developpement-durable.gouv.fr/programmation-pluriannuelle-de-l-energie-ppe-r336.html (in French)

Visual Analytics for Extracting Trends from Spatio-temporal Data

Michiel Dhont[1,2]($\boxtimes$) (iD), Elena Tsiporkova[1], Tom Tourwé[1],
and Nicolás González-Deleito[1]

[1] Sirris, Bd. A. Reyerslaan 80, 1030 Brussels, Belgium
{michiel.dhont,elena.tsiporkova,tom.tourwe,nicolas.gonzalez}@sirris.be
[2] Department of Electronics and Information Processing (ETRO), VUB,
Brussels, Belgium

Abstract. Visual analytics combines advanced visualisation methods
with intelligent analysis techniques in order to explore large data
sets whose complexity, underlying structure and inherent dynamics are
beyond what traditional visualisation techniques can handle. The ultimate goal is to expose relevant patterns and relationships from the data,
since not everything can be exposed easily through intelligent analysis
techniques. On the contrary, the human eye can outperform algorithms
in grasping and interpreting subtle patterns, provided it is supported by
intelligent visualisations.

In this paper, we propose three novel visual analytics techniques for
analysing spatio-temporal data. First, we present a fingerprinting technique for discovering and rapidly interpreting temporal and recurring
patterns by use of circular heat maps. Next, we present a technique supporting comparisons in time or space by use of circular heat map subtraction. Finally, we propose a technique enabling to characterise and
get insights of the temporal behaviour of the phenomenon under study
by use of label maps.

The potential of the proposed approach to reveal interesting patterns
is demonstrated in a case study using traffic data, originating from multiple inductive loops in the Brussels-Capital Region, Belgium.

Keywords: Visual analytics · Temporal statistical analysis ·
Spatio-temporal clustering · Traffic · Covid-19

1 Introduction

Although the recent advances in AI and in particular deep learning techniques
made the exploitation of large volumes of data widely accessible, their black box
nature does not offer an intuitive way of getting a deeper understanding of the
underlying mechanisms of the phenomenon under study. This is usually pursued

This research was subsidised by the Brussels-Capital Region - Innoviris and received
funding from the Flemish Government (AI Research Program).

© Springer Nature Switzerland AG 2020
V. Lemaire et al. (Eds.): AALTD 2020, LNAI 12588, pp. 122–137, 2020.
https://doi.org/10.1007/978-3-030-65742-0_9

via statistical analysis and visualisation techniques, such as correlation analysis, histograms, box- and time-plots, etc. However, such traditional techniques are rather limited when dealing with complex data sets consisting of multivariate spatio-temporal data covering large periods of time.

Visual analytics combines advanced visualisation methods with intelligent analysis techniques in order to explore large data sets whose complexity, underlying structure and inherent dynamics are beyond what traditional visualisation techniques can handle. The ultimate goal is to expose relevant patterns and relationships from the data since not everything can be revealed easily through intelligent analysis techniques. On the contrary, the human eye can outperform algorithms in grasping and interpreting subtle patterns, provided it is supported by intelligent visualisations. Complex data sets need therefore to be manipulated in an intelligent way in order to reveal and highlight the underlying patterns and relationships. This is exactly what visual analytics is about, i.e. advanced data analysis techniques combined with (interactive) visualisation algorithms in order to support the analytical reasoning for decision making.

This paper is concerned with novel visual analytics techniques for spatio-temporal data. First, we present a fingerprinting technique for discovering temporal and recurring patterns by use of circular heat maps. Next, we present a technique supporting comparisons in time or space by use of circular heat map subtraction. Circular heat maps facilitate rapid identification and interpretation of temporal patterns. Finally, we propose a technique enabling to characterise and get insights of the temporal behaviour of the phenomenon under study by use of label maps.

The techniques are application-agnostic and can be used for exploring and extracting relevant insights from spatio-temporal data in domains as traffic, solar and wind energy, electricity consumption, etc. The potential of the proposed approach to reveal interesting patterns is illustrated on recent publicly available traffic count data from the Brussels-Capital Region in Belgium, including the lockdown period imposed by the Covid-19 measures. The latter offers more opportunities for discovering some intriguing trends in the otherwise pretty monotone and typically dense Brussels's traffic.

The rest of the paper is organised as follows. Section 2 presents a literature study of related work, together with a motivation of the rationale of this work. Both the used and novel proposed methods are explained in Sect. 3. Section 4 illustrates the proposed methods on a case study of Brussels traffic data. Finally, Sect. 5 concludes the paper with some possible extensions for further research.

2   Rationale and Related Work

2.1   Rationale

In many application domains, data sets explicitly include a location component, often next to a temporal component. Examples of such spatio-temporal data sets can be found in the renewable energy, mobility or environmental domains. Extracting trends and insights from such data sets by using statistical methods only is not sufficient, as these typically do not exploit the spatial information, nor correlate it with the temporal information. For this reason, visual analytics can serve as the key means to support users in exploring aspects of interest within such data. The use of well-thought visualisations (think of aspects as colours, shapes, positioning, etc.), of suitably processed data (normalisation, aggregation, data imputation, clustering, etc.), explicitly linking the spatial and temporal information, enables the user to derive insights beyond what standard statistical methods can achieve. It optimally supports the competences of the human eye to detect and interpret visual structures, and is hence instrumental in more advanced exploitation of such data, as it facilitates the understanding of the underlying mechanisms of the phenomenon under study.

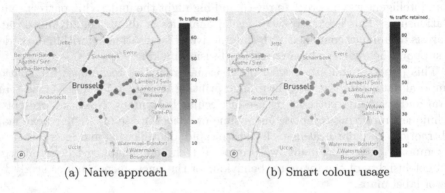

(a) Naive approach   (b) Smart colour usage

Fig. 1. Two ways to visualise retained Brussels traffic by Covid-19 restrictions (Color figure online)

We illustrate this with a real-life example. In January 2020, we started gathering publicly available traffic data captured at 55 locations in the Brussels-Capital Region, Belgium, without having the slightest suspicion that the escalation of the Covid-19 pandemic would deliver interesting data to analyse. The lockdown measures introduced on March 13th imposed serious restrictions on traffic, permitting only limited commuting. As expected, the overall traffic volume reduced dramatically during this period. Figure 1a and Fig. 1b illustrate the percentage of traffic volume retained during the first 4 weeks of the lockdown in comparison to the period before, for all the observed locations. In these figures, the higher the colour intensity of the small circles denoting the different locations, the higher the percentage of regular traffic that was retained for this location. In the naive approach of Fig. 1a, only tints of red are used, which does not reveal much. However, using a colour range with two colours as in Fig. 1b allows to zoom in deeper in the data. In that figure, the border between blue and red is fixed on the mean retained traffic over all the locations on the map. In this way, the blue circles denote locations where a bigger reduction of traffic was observed during the lockdown restrictions. One can now clearly see that the ring road around Brussels's city centre retains proportionally more traffic volume than the residential areas around it. This means that the functional traffic in the Brussels-Capital Region has been less impacted by the Covid-19 restrictions than the recreational traffic. This example illustrates that even basic data analysis can provide more value to existing visualisation methods.

2.2 Related Work

Visual Analytics. The availability of large amounts of data and the ability to analyse and understand it is becoming ever more relevant and important. By automatically exposing the underlying information, via advanced data analysis, one is able to take much more informed decisions. This is useful in some domains, such as marketing, but vital in the field of medical research and political decisions. For instance, the National Visualisation and Analytics Centre (NVAC) from the United States has the mission to use next-generation technologies to reduce the risk of terrorism. In 2006, NVAC assembled a panel of about 40 leading experts from government agencies, industry and academia to outline an R&D agenda [14] with the explicit goal to advance the state of the science to enable analysts to detect the expected and discover the unexpected from massive and dynamic information streams and databases consisting of data of multiple types and from multiple sources, even though the data is often conflicting and incomplete. The agenda also defined visual analytics as a multidisciplinary field that includes the following focus areas: (i) analytical reasoning techniques, (ii) visual representations and interaction techniques, (iii) data representations and transformations, (iv) techniques to support production, presentation, and dissemination of analytical results.

Statistical Visualisations. Data visualisation is typically concerned with depicting some statistics. There is a wide range of classical approaches available, going from very basic ones (e.g. scatter plots) to complex multi-plot visualisations. Depending on the application context and the data analytics workflow considered, suitable visualisations can be selected [17]. However, researchers should not refrain from experimenting beyond the traditional approaches by combining multiple known visualisations into one comprehensive plot. For instance, Allen et al. [1] proposed to augment a violin plot with scatter plots and similar statistics as in a box plot. Furthermore, they proposed possible extensions on their so called rainbow plots by changing orientations and dividing the data in separate groups. These plots succeed in bringing many pieces of information together in an orderly manner in one visual. Another example is shown in the research of Zhao et al. [18], where the advantage of integrating classical line charts into circular heat maps is illustrated.

Constructing appropriate visualisations can be hard and time consuming. For this reason, research has been conducted on constructing powerful recommendation tools which guide users into obtaining relevant visualisations for their specific analytical task [8,16]. Although promising results have been obtained, the main difficulty resides in transferring the specific intent of the research to the recommendation tool.

Visualisation of Spatio-Temporal Data. Sensors which capture data at a fixed frequency are omnipresent. To exploit the (multivariate) time-sensitive data they generate, the time aspect is often treated similar as the sensor measures themselves. However, time requires a special treatment since it is not simply a measure. For this purpose, a range of special-purpose time-sensitive visualisation techniques have been developed and proven to be effective. To decide which technique is useful in a certain situation, one needs to consider aspects as whether the data is dynamic, consists out of events, is multivariate, etc. [11]

Real-world time series data is often enriched with position information, but visualising such data is hard, e.g. visualising dynamically changing data across different geographical locations. Rodrigues et al. [10] proposed a basic two-tier interface to tackle such challenges, which they validated on data concerning energy production of power plants. In the first tier, users see a geographical map indicating information with only few details. To access the second tier, users can select a location, resulting in charts of the energy production.

Spatio-temporal data can also be found in the mobility domain. Tang et al. [13] proposed a method to extract, by an interactive visual analysis system, characteristics on specific areas based on GPS data originating from taxis. They used maps to visualise the main traffic flows and heat maps to observe traffic distribution over time. The resulting visualised characteristics can e.g. assist the business development process in choosing locations for new stores.

In [18], Zhao et al. illustrated the convenience of multiple variants on circular heat maps in spatio-temporal data sets. Sun et al. [12] developed a method to embed spatio-temporal information in a map, by the use of on bidirectional line charts on road sections. Andrienko and Andrienko [2] investigated aggregation strategies in case of spatio-temporal data for both traffic-oriented and movement-oriented visualisations. As visualisations, they proposed multiple variants of directed graphs and heat maps.

3　Methods

This section describes the methods used and proposed in this paper. First of all, Sect. 3.1 describes how cluster analysis is typically performed, as clustering is used in Sect. 3.4 and later in Sect. 4. The remaining subsections describe the visual analytics methods proposed in this paper: temporal fingerprinting through circular heat maps (Sect. 3.2), spatio-temporal comparison through circular heat map subtraction (Sect. 3.3), and temporal behaviour characterisation through label maps (Sect. 3.4).

3.1　Cluster Analysis

Clustering approaches are often used in data science to gain valuable insights by observing which data objects are grouped together. To divide data objects into disjoint clusters, the most commonly used partitioning algorithms require that the number of clusters (k) is determined, either beforehand or when determining the best cut in the dendrogram in case of hierarchical clustering [9]. This represents a challenge, since there is often a lack of prior knowledge to decide this number. Determining a correct, or suitable, k is a hard problem in a real-world data set. To address this issue, researchers usually generate clustering results for multiple values of k, and subsequently assess the quality of the obtained clustering solutions. In situations where no prior knowledge is available, assessing the quality of these solutions can be done using several measures, related to e.g. the compactness and separation properties of the solution (*Davis-Bouldin Index* [4]), the connectedness (*Connectivity* [5]), or the ratio of the within-cluster variance with the overall-between cluster variance (*Calinski-Harabasz Index* [3]). In practice, a majority voting approach is often used, combining the results of multiple such validation measures to identify the most optimal number of clusters.

3.2 Temporal Fingerprinting Through Circular Heat Maps

The analysis of temporal data can often benefit from comparison between recurring time periods as days, weeks, months, etc. which allows to identify trends and seasonality. This is particularly relevant for applications where the monitored phenomenon can be naturally divided in such periods e.g. periodic electricity consumption of households, traffic intensity, yield of photovoltaic (PV) plants, production efficiency of different shifts in a factory, etc.

In order to visualise such trends and patterns explicitly, we have developed a general methodology allowing to convert time series data into a series of circular heat maps covering recurring time periods. Each heat map can be interpreted as a characteristic fingerprint facilitating rapid perception of the behaviour of the phenomenon under study for the time period covered by the heat map. The choice of a circular heat map, instead of a simple line chart or a classical rectangular heat map for example, is motivated by its ability of depicting several dimensions or views therein (i.e. days of the week, hours of the day, vehicle counts) in a visually very compact fashion. This compact representation enables a viewer to quickly find patterns in the data without requiring to focus on different potentially far apart points in the figure. Its circular nature also enables to easily highlight patterns occurring at the limits of the circular dimension (e.g. around midnight). Furthermore, through the use of the *small multiples* visualisation technique [15]—a set of similar thumbnail-sized figures which represent the same phenomenon along a different partitioning of the data—, they facilitate comparisons and the highlight of differences. More precisely, by constructing a small multiple from a collection of fingerprints one is able to do a comparison between fingerprints in different time periods or other multiple phenomena (e.g. occurring in different spatial locations).

In Fig. 2a, such a circular heat map small multiple is generated for depicting the electricity consumption of a university building in Arizona for 5 consecutive months. In this example, each circular heat map depicts the days of the week as concentric circles starting with Monday in the inner circle, followed by Tuesday in the next circle and so on until placing Sunday in the outermost circle. The circles are divided in 24 sectors of 1 h, ordered clockwise and starting with

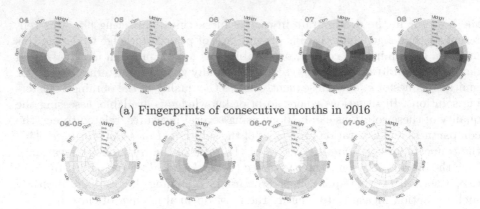

(a) Fingerprints of consecutive months in 2016

(b) Difference between consecutive months (red: increase, green: decrease)

Fig. 2. Energy consumption fingerprints for a university building (Data source: energy consumption at Arizona State University (ASU), https://www.kaggle.com/pdnartreb/asu-buildings-energy-consumption/activity)

midnight at 12 am. The colour of a sector indicates the observed consumption, the darker the colour the higher the observed consumption. In the leftmost fingerprint (April), one can observe that the highest energy consumption occurs between 4 am and 7 pm during weekdays. Thanks to the small multiple, we can observe that the overall consumption pattern is consistent across the months, but the actual consumption increases steadily.

3.3 Spatio-temporal Comparison Through Circular Heat Map Subtraction

Detecting Differences in Time. The use of circular heat maps and small multiples as illustrated in Fig. 2a is well-suited for monitoring evolution in time. One can go even further and subtract the values from two heat maps in order to reveal and highlight better their differences. Subsequently, a heat map with one colour (e.g. white) representing identical values and two diverging colours representing the positive and negative differences can offer a very insightful view.

Depending on the application context, two different subtracting approaches can be considered. For both of them, let us consider a sequence of heat maps covering the same time duration (e.g. a week or a month).

- Compare each heat map in the sequence with the heat map from the **previous period** (by subtracting the latter from the former). For this, it is essential that the heat maps are ordered chronologically in time. In this way, by explicitly highlighting the differences between consecutive time periods, one can more clearly observe local changes. For instance, the fingerprints of Fig. 2a are subtracted from each other, resulting in Fig. 2b. Note that the clear increase of daily consumption between May, June and July and the slight decrease between July and August can be spotted immediately in Fig. 2b, which is much less obvious in Fig. 2a.

- Compare each heat map in the sequence with a **reference (baseline) heat map** (again, by subtracting the latter from the former) to detect global differences in performance. In this way, by examining the resulting sequence of multiple subtracted heat maps from different time periods (e.g. each week), one can quickly identify which period deviates the most from the expected pattern. Figure 3a shows as baseline the average weekly energy consumption pattern of the university building over the full time period. Figure 3b provides the quarterly differences w.r.t. this baseline as a sequence of subtracted heat maps. The observed increase in energy during Q2 and Q3 could be due to the use of air conditioning during warmer months.

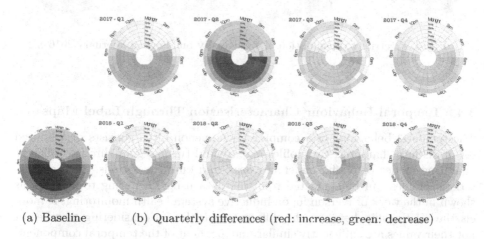

(a) Baseline   (b) Quarterly differences (red: increase, green: decrease)

Fig. 3. Quarterly differences of energy consumption, compared to the overall averaged baseline, for a university building

Detecting Differences in Space. The previous examples strongly emphasise the time aspect. However, our fingerprinting approach can also be applied for analysing data along the spatial component. For example, it is possible to capture temporal patterns (e.g. weekly electricity consumption) in a representative heat map and subsequently, link multiple locations together into the spatial dimension by constructing a sequence of heat maps covering the same time period for all the considered locations. Note that this is another example of the use of small multiples, and many different variables could be used to construct it (e.g. PV power production per weather condition).

Figure 4 provides the fingerprints of the electricity consumption in February 2016 for 4 different buildings of Arizona State University[1]. One can clearly

---

[1] Data source: energy consumption at Arizona State University (ASU), https://www. kaggle.com/pdnartreb/asu-buildings-energy-consumption/activity.

observe 4 quite different consumption patterns: a more intense consumption occurring during the day on weekdays, esp. during afternoons, (heat map labelled with an A); a higher consumption during evenings, both during weekdays and weekends (heat map B); a slightly higher consumption at night and on Mondays (heat map C); and a high consumption spread over all days and hours, but more intense during evenings and at night (heat map D).

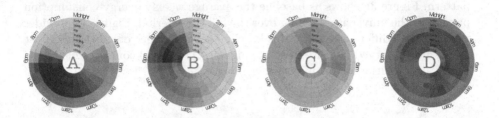

Fig. 4. Electricity consumption of 4 university buildings in February 2016

3.4   Temporal Behaviour Characterisation Through Label Maps

The temporal behaviour of a complex phenomenon often consists of a limited set of distinct characteristic profiles, e.g. a wind turbine goes through different operating modes or traffic undergoes peak and off-peak periods. Such characteristic profiles can be extracted from the data using clustering techniques, as shown in the work of Iverson [6] on inductive system health monitoring. It models the relationship between the different variables by considering whether or not their values are sufficiently similar, independent of the temporal component. This results in a limited number of groups that characterise regular but different behaviour. As a result, each timestamp is assigned a particular label that corresponds to this characteristic behaviour profile.

In a second step, such regular behaviour profiles can be used for different purposes e.g. rapid annotation, detecting deviations, understanding state transitions in time, etc. To this extent, we propose to visualise the resulting behaviour characterisation using a label map, a matrix-like visualisation where each column represents a fine-grained view on the time dimension (e.g. a timestamp such as hour of the day), each row a coarser-grained view on time encompassing all the columns of the row (e.g. a day), and each cell is painted with the colour of the profile to which each timestamp belongs. The choice of such a visualisation, instead of a line chart or a circular heat map for example, is motivated by its ability of clearly depicting behaviour expected to be seen in a recurrent fashion (e.g. every day) on the different rows of the matrix. Furthermore, by just adding new rows, the visualisation can be directly used for real-time monitoring.

Figure 5 illustrates a label map in the previously introduced context of the electricity consumption data of a university building. In this visualisation, columns depict timestamps during one day and rows depict individual days.

Through clustering, five behaviour profiles have been identified. By looking at when the clusters occur, we can observe a (to be expected) clear repetitive weekly pattern, consisting of low consumption during night and weekend (blue/E), high consumption during working hours (yellow/B), and three other profiles most probably related to maintenance operations. The visualisation reveals a change starting on July 18th, where the extent of cluster B (yellow) is reduced; this might be due to the start of the summer holidays.

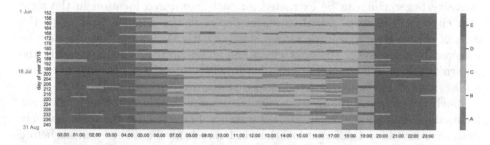

Fig. 5. Evolution of electricity consumption modes in a university building (Color figure online)

4   Case Study on Brussels Traffic

In this section we illustrate how the proposed methods from Sect. 3 can be used to derive relevant insights from spatio-temporal data.

4.1   Data

Our analysis is based on open traffic data from the Brussels-Capital Region, Belgium. We started gathering the data in the beginning of 2020. The lockdown measures introduced on March 13th imposed serious restrictions on the traffic in Brussels, permitting only limited commuting. In this way, the observed traffic after the introduction of the measures can be considered as an opportunity to derive some characteristic blueprints of the traffic in Brussels. Since the second half of May, the Covid-19 restrictions were gradually being relaxed, traffic was slowly returning to 'normal' and allowed us to observe the emergence of traffic volumes associated to different activities.

The data contains vehicle counts, average speed measurements and occupancy (percentage of road covered) of 55 busy locations in Brussels. The locations can be observed in Fig. 1. Each location represents one direction of a road and combines the information of all available lanes in that direction. The obtained data has a one minute granularity and originates from both ANPR cameras and inductive loops. We collected the data in real-time from a publicly available API of Brussels Mobility[2] over a time span from mid January 2020 until the first week of June 2020.

4.2 Unravelling Volume Patterns of Brussels Traffic

Weekly Traffic Intensity Patterns. City traffic is strongly dependent on the day of the week. Therefore, the fingerprinting approach proposed in Sect. 3.2 can be applied by segmenting the data per week. In this way, a weekly traffic intensity fingerprint can be extracted for each monitored location in the form of a circular heat map. Like before, we depict in the circular heat maps the days of the week as concentric circles starting with Monday in the inner circle, followed by Tuesday in the next circle and so on until placing Sunday in the outermost circle. The circles are divided in 24 sectors of 1 h (i.e. vehicle counts are aggregated per hour), ordered clockwise starting with midnight at 12 am.

Such a representation allows to easily compare weekly patterns of different locations for different time periods. For instance, Fig. 6 depicts the typical weekly patterns derived for 4 different locations in Brussels for 2 different periods before and during the Covid-19 lockdown (i.e. the corresponding weeks are aggregated into one weekly pattern per location). The same colour scale is used for all the four locations, which facilitates objective comparison across them. It is clear that Troontunnel has more traffic than Belliardtunnel, which has on its turn more traffic than Keizer Karellaan and Vleurgattunnel. This pattern is apparent before as well as during the lockdown. It is also interesting to zoom into the specificity of the weekly traffic behaviour. For instance, despite the well manifested difference in the overall traffic intensity between Vleurgattunnel and Belliardtunnel, a very clear morning peak (dark sector between 7 and 9 am) can be detected for both locations during the working days in the pre-Covid-19 period, while the evening peak (dark sector between 4 and 6 pm) is well established only for Belliardtunnel. During the lockdown, this morning peak during working days is still observable (except for Vleurgattunnel), indicating that some work commuting was still happening through those locations. The third row of the figure depicts the difference between the pre-Covid-19 situation of the top row and the full lockdown situation of the second row. This allows us to observe that night traffic during the weekend days has disappeared completely (esp. visible for the Troontunnel and the Belliardtunnel), that the weekend days had a larger reduction in traffic compared to weekdays for the Troontunnel, and that the reduction was also stronger for rush hours in both Troontunnel and Belliardtunnel.

It is also interesting to study the traffic intensity evolution. In Fig. 7, characteristic weekly fingerprints are depicted based on aggregating vehicle counts over all available locations from the 10th until the 15th week of 2020. The week number of each fingerprint is shown in its left upper corner. By examining these fingerprints one can detect easily the introduction of the first Covid-19 restrictions on the Saturday of week 11, and the introduction of the full lockdown on

---

[2] https://data-mobility.brussels/traffic/api/counts/.

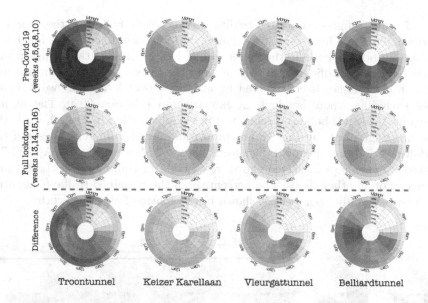

Fig. 6. Vehicle counts before and during the Covid-19 restrictions at 4 locations

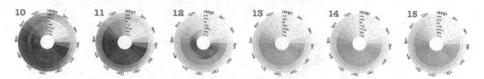

Fig. 7. Weekly evolution of vehicle counts in Brussels (week 10 until week 15)

Wednesday at 12 am in week 12. Moreover, by comparing week 10 (pre-Covid-19) with weeks 13, 14 and 15 (full lockdown), it is interesting to observe that the biggest reduction in traffic is obtained during the weekend. The intensity of colours during weeks 13, 14 and 15 is also very similar, indicating that people consistently obeyed the imposed restrictions.

Traffic Volume Disaggregation. The collected data spans over a time period covering 3 distinct traffic situations: 1) normal traffic referring to regular work-school weeks; 2) carnival holidays referring to the school vacation in week 9, which excludes school-related traffic and some work-related traffic due to parents taking vacation—at the same time, those families have more time for recreational trips (e.g. city trips, sport, shopping, etc.); and 3) lockdown weeks referring to the period of activity restrictions due to the Covid-19 measures, including only work related traffic which cannot be performed via teleworking and other minimal essential traffic (e.g. shopping for food).

The Covid-19 restrictions were gradually relaxed since the second half of May. Traffic was slowly returning to 'normal' and allowed us to observe the emergence of traffic volumes associated to different activities. Following the approach described above, we have generated characteristic weekly fingerprints for different time periods depicted in the upper row of Fig. 8. Each of these fingerprints depicts the hourly traffic intensity throughout the (averaged) week. Our baseline (100% traffic volume) is constructed by averaging over 5 'regular' work-school weeks, excluding school holidays, in January and February 2020. The second fingerprint (Carnival holidays) refers to the school vacation in week 9, while the remaining fingerprints average traffic intensities over the different phases of the lockdown period, i.e. complete lockdown between March 14th and April 17th, followed by opening of selected shops, re-starting of companies' activities and re-opening of all other shops. The percentage of the remained traffic (compared to the normal traffic) per week is shown in the centre of each fingerprint.

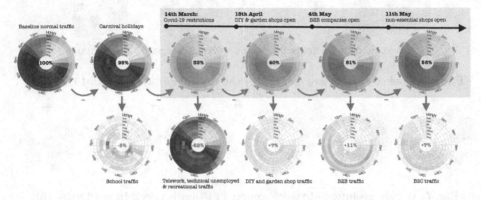

Fig. 8. Upper row: grouped fingerprints of vehicle counts in Brussels. Bottom row: disaggregated fingerprints from above (green: reduction, red: increase) (Color figure online)

Comparing the characteristic fingerprints allows to disaggregate the overall traffic volume into separate intensities associated with different activities. This is realised in the lower part of Fig. 8, which depicts the result of subtracting each weekly fingerprint in the top part of the figure from the fingerprint immediately on its left. This highlights better the time slots where traffic has increased (red) or decreased (green) compared to the previous fingerprint. For instance, since the introduction of the first relaxation measures on April 18th, when people were allowed again to go to do-it-yourself and garden shops, traffic increased on average with 7% (3rd fingerprint on the bottom row). It is remarkable that this increase in traffic can also be observed on Sunday, when shops are closed. This seems to suggest that the traffic might not be exclusively for shopping, but that people perhaps stopped following travel restrictions strictly.

4.3   Insightful Blueprints of Brussels Traffic

The purpose of the following analysis is to get better insights of different traffic modes of the small ring of Brussels, which is actually a sequence of many tunnels crossing Brussels from one side to the other and is notoriously known for frequent traffic problems.

Traffic Profile Extraction. Our analysis focuses on 16 locations of the small ring of Brussels. Our first aim is to identify characteristic traffic modes when considering all the 16 locations as one connected trajectory of tunnels. For this purpose, we perform the following steps:

- We consider only data from the weeks before the Covid-19 restrictions and exclude the carnival holidays. We order the data set in such a way that the locations are sequentially ordered as they spatially appear in reality.
- We associate with each timestamp a vector of 48 dimensions, based on the vehicle count, speed and road occupancy measurements that are available for each location.
- We average for each timestamp the measures over a rolling time window of 10 min, in order to achieve more resilient, but still fine-grained (per minute) results, Since this real-world data contains missing timestamps, this approach enables us to still use an estimation by taking the average of the non-missing values within the 10 min time window.
- We scale each of the measures by the min-max normalisation [7]. This way all values will be between 0 and 1, resulting in equal weights during clustering.
- We cluster these vectors using k-means clustering using the Euclidean distance, resulting in clusters (of timestamps) representing characteristic traffic modes. The number of clusters was determined by majority voting of multiple validation measures, as explained in Sect. 3.1, resulting in 5 clusters.

In order to facilitate the semantic interpretation of the clusters (or traffic modes), we label them as follows: **Mode A:** Night traffic (avg. occupancy 7%); **Mode B:** Morning rush hour peak (avg. occupancy 26%); **Mode C:** Day traffic outside rush hours (avg. occupancy 22%); **Mode D:** Evening rush hour peak (avg. occupancy 28%); **Mode E:** Early morning and late evening traffic (avg. occupancy 15%). Remark that, in practice, occupancy is never close to 100% since this would require all vehicles to touch each other.

Temporal Traffic Blueprints. In a second step, we use the traffic modes previously identified and visualise them in a label map. Figure 9 shows a label map as proposed in Sect. 3.4, which labels each timestamp using the 5 labels identified above. Note that the grey zones in the plot are caused by missing data stretc.hing over more than the 10 min time window.

This visualisation allows to observe the following interesting phenomena:

- Traffic during the first weeks of the lockdown was significantly reduced, resembling low intensity night traffic (mode A).

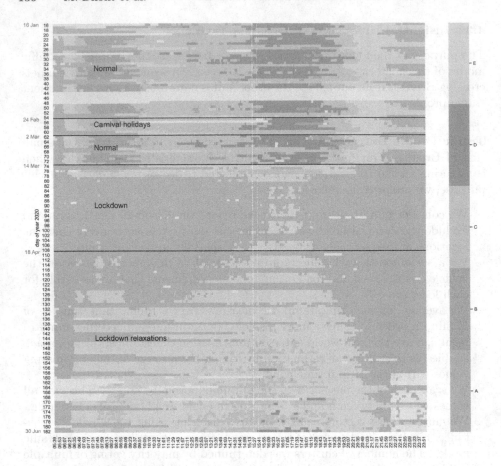

Fig. 9. Evolution of traffic clusters (modes) in Brussels

- During lockdown relaxations, traffic gradually evolved towards what is normally light traffic in the late evening and early morning (mode E)
- Day traffic outside rush hours (mode C) emerges in a similar way as mode E while lockdown measures are relaxed, initially starting at specific periods in the afternoon but slowly unrolling forward covering the whole afternoon.
- The afternoon rush hour (mode D) which can normally be observed between 15:00 and 18:00 completely disappeared during the lockdown and has since (almost) not yet reappeared.

5   Conclusion and Future Work

In this work, we proposed a range of visual analytics methods dedicated to spatiotemporal data. The power of the proposed visualisations lies in the transformation of data analytics results into well-thought visual representations revealing underlying insights. Questions as how to aggregate data, partition data, normalise data, represent it as a feature vector, etc. are the key to arrive at insightful visual representations facilitating the human eye's discovery process.

There are still many opportunities to expand this research, e.g. increase the intelligibility of visuals by making them interactive. Enabling to zoom in on time periods of interest or to change dynamically some parameters might be very valuable in supporting human-in-the-loop data exploration.

References

1. Allen, M., Poggiali, D., Whitaker, K., Marshall, T.R., Kievit, R.: Raincloud plots. PeerJ Preprints **6**, e27137v1 (2018)
2. Andrienko, G., Andrienko, N.: Spatio-temporal aggregation for visual analysis of movements. In: Symposium on Visual Analytics Science and Technology. IEEE (2008)
3. Caliński, T., Harabasz, J.: A dendrite method for cluster analysis. Commun. Stat. Theory Methods **3**(1), 1–27 (1974)
4. Davies, D.L., Bouldin, D.W.: A cluster separation measure. IEEE Trans. Pattern Anal. Mach. Intell. **2**, 224–227 (1979)
5. Handl, J., Knowles, J., Kell, D.B.: Computational cluster validation in postgenomic data analysis. Bioinformatics **21**(15), 3201–3212 (2005)
6. Iverson, D.L.: Inductive system health monitoring. In: NASA (2004)
7. Liu, Z., et al.: A method of SVM with normalization in intrusion detection. Procedia Environ. Sci. **11**, 256–262 (2011)
8. Luo, Y., Qin, X., Tang, N., Li, G.: DeepEye: towards automatic data visualization. In: 34th International Conference on Data Engineering. IEEE (2018)
9. MacQueen, J., et al.: Some methods for classification and analysis of multivariate observations. In: Proceedings of the Fifth Berkeley Symposium on Mathematical Statistics and Probability, Oakland, CA, USA, vol. 1, no. 14, pp. 281–297 (1967)
10. Rodrigues, N., et al.: Visualization of time series data with spatial context. In: Proceedings of the 10th International Symposium on Visual Information Communication and Interaction, pp. 37–44 (2017)
11. Spears, W.M.: An overview of multidimensional visualization techniques. In: Evolutionary Computation Visualization Workshop (1999)
12. Sun, G., Liang, R., Huamin, Q., Yingcai, W.: Embedding spatio-temporal information into maps by route-zooming. IEEE Trans. Visual. Comput. Graph. **23**(5), 1506–1519 (2016)
13. Tang, Y., Sheng, F., Zhang, H., Shi, C., Qin, X., Fan, J.: Visual analysis of traffic data based on topic modeling (ChinaVis 2017). J. Visual. **21**(4), 661–680 (2018)
14. Thomas, J.J., Cook, K.A.: A visual analytics agenda. IEEE Comput. Graph. Appl. **26**(1), 10–13 (2006)
15. Tufte, E.R., Goeler, N.H., Benson, R.: Envisioning Information, vol. 126. Graphics press, Cheshire (1990)
16. Vartak, M., Huang, S., Siddiqui, T., Madden, S., Parameswaran, A.: Towards visualization recommendation systems. ACM SIGMOD Rec. **45**(4), 34–39 (2017)
17. Wong, P.C., Thomas, J.: Visual analytics. IEEE Comput. Graph. Appl. **1**(5), 20–21 (2004)
18. Zhao, J., Forer, P., Harvey, A.S.: Activities, ringmaps and geovisualization of large human movement fields. Inf. Visual. **7**, 198–209 (2008)

Layered Integration Approach for Multi-view Analysis of Temporal Data

Michiel Dhont[1,2]([✉]) [iD], Elena Tsiporkova[1], and Veselka Boeva[3] [iD]

[1] Sirris, Bd A. Reyerslaan 80, 1030 Brussels, Belgium
{michiel.dhont,elena.tsiporkova}@sirris.be
[2] Department of Electronics and Information Processing (ETRO), VUB,
Brussels, Belgium
[3] Blekinge Institute of Technology, Karlskrona, Sweden
veselka.boeva@bth.se

Abstract. In this study, we propose a novel data analysis approach that can be used for multi-view analysis and integration of heterogeneous temporal data originating from multiple sources. The proposed approach consists of several distinctive layers: (i) select a suitable set (view) of parameters in order to identify characteristic behaviour within each individual source (ii) exploit an alternative set (view) of raw parameters (or high-level features) to derive some complementary representations (e.g. related to source performance) of the results obtained in the first layer with the aim to facilitate comparison and mediation across the different sources (iii) integrate those representations in an appropriate way, allowing to trace back similar cross-source performance to certain characteristic behaviour of the individual sources.

The validity and the potential of the proposed approach has been demonstrated on a real-world dataset of a fleet of wind turbines.

Keywords: Data integration · Data mining · Temporal data clustering · Multi-view learning

1 Introduction

Mining data collected from continuous monitoring of industrial assets in the field allows to derive relevant insights about their operations and performance. Such complex real-world datasets are usually composed of heterogeneous subsets (or multi-views) of parameters, which should be considered explicitly during analysis in order to exploit fully the richness of the data. For instance, the performance of an industrial asset is impacted by a diverse set of factors e.g. operating modes concerned with the internal working of the asset and exogeneous factors such as weather conditions. However, it is not trivial to directly link or trace back certain

This research was subsidised by the Brussels-Capital Region - Innoviris, received funding from the Flemish Government (AI Research Program) and was supported by the Energy Transition Fund of the FPS Economy through the project BitWind.

© Springer Nature Switzerland AG 2020
V. Lemaire et al. (Eds.): AALTD 2020, LNAI 12588, pp. 138–154, 2020.
https://doi.org/10.1007/978-3-030-65742-0_10

performance to distinct operating modes due to the multitude of influencing factors, which are often also highly interdependent.

In addition, real-world datasets often originate from different sources, which may differ in period coverage, resolution, data quality, technical configuration, etc. Pooling multi-source datasets together, which is often done to increase statistical representativeness, requires standardization and normalization, which often leads to information loss and may mask source-specific features. For instance, mining for distinct operating modes is more appropriate to be pursued per asset, rather than pooling everything together, since not all assets might go through all operating modes. This implies that one might need to approach multi-source analysis in an incremental fashion rather than aiming for brute force integration of all the available data.

Classical data mining and analysis approaches have still some shortcomings in this aspect aiming at delivering a total integration solution at once. An alternative approach is to **exploit the multi-view nature of the data**. Some rewarding techniques of multi-view mining have been already proposed in the literature [1,14]. However, they all were concerned with single-source datasets and dedicated to one specific mining approach (e.g. clustering, deep learning or classification). This research provides a general analysis methodology, which is agnostic to the specific mining techniques used and focuses on the following key aspects: initial *individual analysis* per source in order to preserve the richness and the authenticity of each source; individual *mediation analysis* per source aiming at bringing the sources closer together; cross-source *integration analysis* aiming at leveraging analysis results across the sources without compromising their individual characteristics.

More concretely, the proposed approach consists of several distinctive layers: (i) select a suitable set (view) of parameters in order to identify characteristic behaviour within each individual source (ii) exploit an alternative set (view) of raw parameters (or high-level features) to derive some complementary representations (e.g. related to source performance) of the results obtained in the first layer with the aim to facilitate comparison and mediation across the different sources (iii) integrate those representations in an appropriate way, allowing to trace back similar cross-source performance to certain characteristic behaviour of the individual sources.

The validity and the potential of the proposed approach have been demonstrated on a real-world dataset of a fleet of wind turbines. We have been able to identify distinctive profiles of production performance and subsequently, have been able to establish an explicit link between those performance profiles and well characterised operating modes.

The rest of the paper is organised as follows. Section 2 reviews related work and discusses the rationale motivating the proposed approach. Section 4 introduces the used methods and formally describes the proposed layered integration approach. Data and experimental setting used for the evaluation purposes are explained in Sect. 5. Section 6 presents the evaluation of the proposed approach and discusses the obtained results. Section 7 is devoted to conclusions and future work.

2   Related Work and Rationale

Multi-view datasets consist of multiple data representations or views, where each one may contain several features [5]. There are many scenarios where data can be described from multiple views [14]. In such multi-view scenarios it is more interesting to consider the diversity of different views rather than simply concatenating them. Furthermore, remote sensor technologies are very accessible these days, resulting in the appearance of high frequency sensor data collected for all kinds of environments and assets. Despite the accelerated development of mining techniques for multi-source data, managing and interpreting multi-source data is still very challenging [15].

One way to exploit multi-source data is by data integration. Data integration is the combination of data from distinct data sources into a meaningful and useful format. It can either aim to bring data together for the purpose of visualization or fuse them together in one integrated dataset. Three main approaches have been developed [6]: (i) *Schema mapping*: a global mediating schema is used, e.g. by defining mappings between the distinct schemas of each data source; (ii) *Record linkage*: records that refer to the same entry across distinct data sources are matched together; (iii) *Data fusion*: data from distinct data sources are combined by probabilistic algorithms. One major risk in constructing an integrated dataset is the risk on losing source-specific characteristics.

2.1   Challenges Related to Real-World Datasets

In this research, we consider real-world datasets originating from multiple data sources, e.g. fleet data. An asset within the fleet captures data from multiple sensors and each sensor can moreover have a different accuracy and reliability. Two main issues arise when one wants to mine such complex real-world datasets.

First of all, exploiting fully all the properties of the captured data is not trivial since it is composed out of several **heterogeneous subsets of parameters**. Consider data generated by wind turbines, consisting of sensor data of operational parameters, such as oil temperature and rotor speed, on the one hand, and data about power production in function of different exogenous factors such as wind speed and outside temperature, on the other hand. Mining such data considering all the parameters at once is often not the best thing to do since the operational parameters are typically analysed in time, while the power production is better monitored as a function of the weather conditions.

In addition, taking into account and combining the information from the **different sources**, such as the fleet of turbines, is far from trivial. Each source may differ in period coverage, resolution, data quality, technical configuration, etc. To optimally use all information one could pool all multi-source datasets together. However, this requires suitable standardisation and normalisation, which could lead to information loss and may mask source-specific parameters. As example one may want to cluster timestamps according to their behaviour in case of wind turbines. However, rather than pooling everything together it is more appropriate to do that per turbine since not all turbines might go through all operating modes and pooling data would lead to noise/sub-optimal clusters.

2.2 Multi-view Learning

Multi-view learning is a semi-supervised approach that aims to obtain better performance by using the relationship between different views [14]. Multi-view unsupervised learning and specifically multi-view clustering has attracted great attention recently due to availability of inexpensive unlabelled data in many application domains [5]. The goal of multi-view clustering is to find groups of similar objects based on multiple data representations. In the past, multi-view clustering approaches have shown to outperform the single-view clustering approach in case of true single-source multi-view datasets. A multi-view clustering approach uses a conditional independence assumption of the different views [1]. However, a perfect conditional independence of different views is almost impossible in real-world datasets. Fortunately, in [7] one illustrates that in a more realistic case where each group (layer) of parameters is not perfectly independent, a similar approach can also be applied to outperform single-view clustering. The latter is called multi-layer clustering. However, a point of attention in those hierarchical clustering approaches is the tendency to construct too small clusters [1].

Hierarchical approaches are not only advantageous in cluster tasks, but can be used in all kinds of data mining strategies. In [14], a comparison is made concerning multiple multi-view learning techniques. The authors' main conclusion is that multi-view learning is effective and promising in practice, but there is still a lot of work to be done to make them useful in a wide variety of applications.

In this paper we propose a multi-layer data analysis methodology which cleverly benefits from the multi-view approach and demonstrates its potential to deal with multi-source data when applied in a well designed incremental fashion.

3 Use Case Context and Ambition

The proposed layered integration methodology is demonstrated on public sensor data originating from a fleet of wind turbines. The initial ambition of the studied experimental scenario is to identify and characterise potentially different operating modes across the fleet. Notice that wind turbines can have several different operating modes, e.g. working at full speed, reduced speed in order to limit the noise burden on the surroundings, tailored production due to oversupply on the net and others. Subsequently, the ultimate goal is to derive distinctive profiles of production performance and establish an explicit link between those performance profiles and the characterised operating modes.

Two main types of input data sources are used to capture the operation of a wind turbine: operational (endogeneous) and environmental (exogeneous) parameters. The former are referring to sensors measuring the internal working of the turbine, such as oil temperature and rotor speed, while the latter are considering different exogeneous factors impacting the production, such as wind speed and temperature. The performance of a wind turbine is typically expressed in terms of the produced active power as a function of the wind speed, called power curve and visualised as depicted in Fig. 2b. A power curve typically has an S shape. Based on this curve, one can derive roughly the expected active power based on a certain wind speed. It is not trivial/possible to determine whether a particular production performance is as expected or there is some deviation since the impact of the internal working of the turbine is not explicitly considered. The same active power output may be induced by different operating modes of the turbine given the same exogeneous context.

The ultimate goal of our analysis is to derive an explicit link between the internal working modes (different compositions of the endogenous parameters) and the expected output (active power) at fleet level. This will enable for quantitative labelling of the turbine operation with respect to the whole fleet, e.g. "as the rest of the fleet", "under-performing", "better than the fleet".

4   Methods and Proposed Approach

4.1   Clustering Analysis

Three partitioning algorithms are commonly used for data analysis to divide the data objects into k disjoint clusters [10]: k-means, k-medians, and k-medoids clustering. The three partitioning methods differ in how the cluster center is defined. In k-means clustering, the cluster center is defined as the mean data vector averaged over all objects in the cluster. In k-medians, the median is calculated for each dimension in the data vector to create the centroid. Finally, in k-medoids clustering, the cluster center is defined as the object with the smallest sum of distances to all other objects in the cluster.

The partitioning algorithms contain the number of clusters (k) as a parameter and their major drawback is the lack of prior knowledge for that number to construct. Unfortunately, determining a correct, or suitable, k is a difficult problem in a real-world dataset. For such cases, researchers usually generate clustering results for different numbers of clusters, and subsequently assess the quality of the obtained clustering solutions.

In the context of the presented study, we have no prior knowledge about the underlying structure of the data. Thus, we use four internal validation measures for analyzing the data and select the optimal clustering scheme. We have selected two validation measure for assessing compactness and separation properties - *Silhouette Index* [11] and *Davis-Bouldin Index* [4], one for assessing connectedness - *Connectivity* [8], and one for assessing the ratio of the within-cluster variance with the overall-between cluster variance - *Calinski Harabasz Index* [3].

4.2   Kernel Density Estimation (KDE)

As the name suggests KDE is a non-parametric method to estimate the probability density function of a random variable density by use of a kernel. Practically, the KDE f'_b is constructed by averaging the sum of a density estimation for each sample $X_1, X_2...X_n$, as shown in Eq. (1). In this formula, K is a kernel function of choice, which needs to be symmetric around zero. Often one uses a Gaussian kernel (see Eq. (2)) [12].

$$f'_b(x) = \frac{1}{nb} \sum_{i=1}^{n} K\left(\frac{x - X_i}{b}\right) \quad (1) \quad K(y) = \frac{1}{\sqrt{2\pi}} \exp\left(-\frac{y^2}{2}\right) \quad (2)$$

In Eq. (1), the hyperparameter bandwidth b acts as a smoothing factor. A large b will spread the kernel function, resulting in a very smooth KDE. However, if b gets too large, a lot of information is smoothed out. Since the ground truth is often unknown, some rules of thumb have been developed in the past. Amongst others, Silverman's rule is often used. This rule is defined as $b = (n(d + 2)/4)^{-1/(d+4)}$, with d the number of dimensions [13].

4.3 Hypercube Binning Approach

The hypercube approach is a method to characterise (discretize) data by a multi dimensional binning approach. A hypercube is defined as a cube of N dimensions. Hypercube binning can be very useful when analysing multi-dimensional data since by dividing the parameter space into cubes, one can derive properties of interest for each cube. These properties might be for example the median, standard deviation or even the KDE of a (not yet used) parameter. The assumption is that the data points characterized with similar parameter values (so they end up in the same hypercube), exhibit similar properties.

4.4 Layered Multi-view Analysis: General Approach

In this study, we propose a novel approach for analysing complex real-world time series data. It is inspired by some previous study of Boeva et al. [2] dealing with the analysis of high-dimensional multivariate data generated in several different experiments. We have conceived a more generic approach, based on the idea that different in nature data parameters form distinctive views of the data and should be considered for separate analysis in a multilayered fashion.

Suppose that a particular phenomenon (e.g. biological/chemical process, physical asset, etc.) is monitored in time via multiple data capturing measurements of different nature (e.g. experimental setup, machine configuration, high-throughput measurements, operational parameters, exogeneous factors, etc.). This will result in collecting measurements of several parameters that each contains part of the relevant information. Furthermore, data analysis can often benefit from considering (pooling together) data from multiple observations/sources of the phenomenon under study, e.g. in case of biological or chemical processes multiple datasets generated in different experimental conditions are frequently explored together, while in industrial contexts datasets originating from a portfolio or a fleet of industrial assets are often consolidated for analysis.

Subsequently, let us assume that we have access to data of N different sources (e.g. a fleet of wind turbines) of the phenomenon under study monitored via n different types of parameters, which are the same across the different observations/sources, while the time periods covered, the data quality and the capturing resolution are not necessary the same and may vary across the sources.

Formally, the main steps (layers) of the proposed multi-view data analysis approach are explained in the subsections below. The overall data corpus is composed of N different datasets (multi-variate time series) $D_1, D_2, \ldots, D_N$, one per source i $(i = 1, 2, \ldots, N)$. Each individual dataset is composed of n time series $D_i = \{D_{i1}, D_{i2}, \ldots, D_{in}\}$, one per monitored parameter.

Individual Analysis Layer (View 1). This layer is concerned with individual per source data analysis, focusing on a subset of relevant parameters allowing to drill down for insights without the necessity to compromise across all sources.

(a) Select a subset of p common in nature parameters across the different sources based on the following criteria:
   - the selected subset of parameters provides *comprehensive view* about a particular aspect(s) (e.g. behavioural, operational or other characteristics) of the studied phenomenon
   - it is feasible to pool together per individual source the corresponding time series for analysis (e.g. cover the same time window and have the same resolution per observation).
(b) For each source i, the corresponding time series $D_{ij_1}, D_{ij_2}, \ldots, D_{ij_p}$, one per monitored parameter j, $(j = 1, 2, \ldots, p)$, are subsequently integrated into a dataset D_{i_p} of dimensions p by t_i (the size of the covered time window per source i), $(i = 1, 2, \ldots, N)$.
(c) Subsequently, each matrix D_{i_p} per source i, $(i = 1, 2, \ldots, N)$ is individually subjected to a suitable analysis (e.g. clustering, regression or classification).
(d) Thus, for each source i, $(i = 1, 2, \ldots, N)$, the foregoing data analysis step has generated a set of results or data models (e.g. clusters or regression functions) $R_{i1}, R_{i2}, \ldots, R_{ik_i}$, where k_i is a source specific parameter.

Mediation Analysis Layer (View 2). This analysis layer is building upon the results from the previous layer by considering an alternative subset of parameters (view) allowing to derive comparative insights across the sources.

(a) Select a subset of q parameters across all sources based on the criteria:
   - the parameters offer an alternative *complementary view* (representation) of the results obtained per source in the individual analysis layer
   - the obtained complementary representations allow for follow up comparative analysis across the different sources.
(b) For each source i, the corresponding time series $D_{ij_1}, D_{ij_2}, \ldots, D_{ij_q}$, one per selected parameter j, $(j = 1, 2, \ldots, q)$, are subsequently joined together to construct a complementary dataset CD_{il_i} for each result R_{il_i}, $(l_i = 1, 2, \ldots, k_i, i = 1, 2, \ldots, N)$.

(c) Subsequently, each complementary dataset CD_{il_i} per source i, is subjected to a suitable further analyse (e.g. profiling or clustering) leading to complementary results CR_{il_i}, ($l_i = 1, 2, \ldots, k_i$, $i = 1, 2, \ldots, N$). The latter can be easily interpreted and compared across the different sources and are uniquely associated with the corresponding results obtained from the previous layer.

Integration Analysis Layer (Linking the Views). This analysis layer is concerned with leveraging the results obtained in the previous analysis layers across the different sources. The ultimate goal is to derive an explicit link between the results generated in the different views.

(a) The results, obtained for each source in the mediation layer, are pooled together, i.e., the following dataset is composed CR_{il_i}, ($i = 1, 2, \ldots, N$, $l_i = 1, 2, \ldots, k_i$) and subjected to consolidation, e.g. grouping similar results. In this way a cross-source integration is achieved delivering a smaller number of representative, across the different sources, results S_r ($r = 1, \ldots, m$) where $m \leq k_1 + \ldots + k_i$ since each S_r is derived from a subset of CR_{il_i}.
(b) Subsequently, for each S_r ($r = 1, \ldots, m$) a unique link can be established with different subsets of the initial results obtained in the very first individual analysis layer i.e. R_{il_i}, ($i = 1, 2, \ldots, N$, $l_i = 1, 2, \ldots, k_i$). For instance, S_r can potentially define some unique representations or labels of distinctive classes formed by the corresponding R_{il_i} subsets.

4.5   Layered Multi-view Analysis: Instantiated in the Use Case

The layered multi-view analysis approach, introduced in Sect. 4.4, is instantiated for the considered fleet of wind turbines use case described in Sect. 3. The overall approach is visualised in Fig. 1.

Recall that, two main types of input data sources are used to capture the operation of a wind turbine: operational (endogeneous) and environmental (exogeneous) parameters. The former are referring to sensors measuring the internal working of the turbine, such as oil temperature and rotor speed, while the latter are considering different exogeneous factors impacting the production, such as wind speed, wind direction and temperature.

Individual Analysis Layer: Operating Mode Characterisation (Internal View). This layer is concerned with data analysis only from the perspective of the internal working of each turbine detached from the other influencing factors i.e. based solely on the operational parameters. The aim is to derive clusters of timestamps with characteristic operating behaviour (operating modes) per turbine. Rather than pooling everything together, it is more appropriate to do that per turbine since it may occur that not all turbines go through all operating modes for the considered time period and pooling data would lead to noise/suboptimal clusters. Moreover, the datasets constructed per turbine may differ in period coverage since considering only the common period coverage may lead to a substantial reduction of the data and also mask some source-specific features.

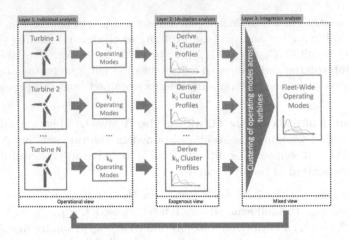

Fig. 1. Layered integration approach for the use case of fleet of wind turbines.

Subsequently for each turbine, a number of clusters are derived grouping together timestamps for which the values of the operational parameters relate to each other in a similar way. It is not necessarily expected that the same number of clusters will be derived for each turbine since as already mentioned above not all turbines go through all operating modes for the considered time periods. The assumption is that each cluster is representing a distinctive operating mode of the turbine. Each cluster will define a range of allowable values for each operational parameter and thus generates parametric characterisation of the mode. In this way, the pool of clusters produced for the fleet leads to the construction of a repository of operating modes as depicted in the left panel of Fig. 1.

Mediation Layer: Performance Profiling (Exogeneous View). In this layer, we pursue a way to derive an alternative representation of each operating mode in terms of expected performance. The richness of our multivariate data allows to consider an alternative view for each cluster of timestamps generated in the previous layer. For instance, it can be useful for monitoring purposes to have an estimation of how likely is to observe certain production output for a given exogeneous context (i.e., wind speed, wind direction and temperature).

Thus for each cluster of timestamps, from the previous layer, a dedicated dataset can be constructed, composed of the corresponding values for wind speed, wind direction, temperature and active power. Such a dataset can be used to derive some performance profile per cluster estimating the expected production of active power. However, the active power behaviour might vary substantially for different exogeneous contexts or in other words for different combinations of the values of the 3 parameters wind speed, wind direction and temperature. Therefore, we will be pursuing the construction of performance profile per cluster in an incremental fashion by using the hypercube binning approach in order to limit as much as possible the impact of the exogeneous factors. The approach is described in more details below:

1. *Hypercube binning* In order to split the active power points into subsets produced in similar context, i.e. exogeneous parameters with similar values, the hypercube binning approach as explained in Sect. 4.3 is applied on each cluster dataset. The number of generated hypercubes depends on the granularity of the binning step. The higher the granularity, the more hypercubes will be constructed per cluster, the less points will be contained at average in each hypercube.
2. *Individual probability distributions per hypercube* As described in the previous step, each hypercube represents a group of similar points from perspective of the exogeneous context. Subsequently, the probability density of the active power can be estimated using the KDE approach from Sect. 4.2
3. *Mixture probability distributions per cluster* The individual distributions derived in the previous step per hypercube in a given cluster are subsequently combined to form mixture distributions for this cluster.

The derived mixture distributions per cluster (see the middle panel of Fig. 1) can be interpreted as distinctive probabilistic profiles of the expected performance in terms of active power produced. It is also important to note that the actual operating mode (the ranges of allowable values for each operational parameter) generating this performance profile can be traced back through the cluster characterisation in the previous layer.

Integration Layer: Fleet-Wide Performance Labeling (Mixed View).
As result of the previous two layers, a repository of operating modes can be constructed, where each operating mode is: 1) characterised in terms of allowable ranges of the operational parameters; 2) associated with a probabilistic profile of expected production. However, the different operating modes have been derived by treating the data of each turbine separately, which does not allow for knowledge transfer and model leverage across the fleet. For instance, considering each set of characterised operating modes per turbine separately is much too limiting since some operating modes might not be observed for some turbines for the considered time window. The latter does not exclude that they might occur in the future. Subsequently, not sharing the operating mode characterisations across the fleet might result into too high rate of unseen operating modes per turbine or in other words high rate of false detection of anomalous operation. Moreover, it is also expected that several different operating modes might be exhibiting very similar production performance.

It is interesting to investigate how many distinct classes/profiles of production performance are detectable at fleet level. The associated with each operating mode probabilistic profiles of expected production can be compared directly with each other since they all are probability density functions of the active power. Subsequently, all the profiles are pooled together and subjected to clustering. In this way, several distinctive profiles of production performance are derived across the fleet (see the right panel of Fig. 1) and subsequently, an explicit link between those performance profiles and the characterised operating modes can be established.

5   Dataset and Implementation

The proposed approach has been validated using public SCADA[1] data originating from a wind turbine fleet of Engie, located in La Haute Borne. The dataset contains measurements of a fleet of 4 wind turbines collected with a 10-minute interval for 31 parameters, listed on GitHub. The data is collected between January 2009 and March 2017.

5.1   Data Preprocessing

Eliminating Correlated Parameters. Some of the monitored parameters in the Engie dataset produce values which are highly correlated due to several reasons 1) monitoring the same phenomenon with multiple sensors, e.g. the nacelle of each turbine is equipped with 2 different anemometers both measuring the wind speed; 2) derived parameters, e.g. the measured wind speed by the two nacelle anemometers is used to calculate the average wind speed; 3) internal dependencies between some parameters, e.g. generator speed and generator converter speed. Therefore in order to avoid over-fitting, only one parameter of the correlated parameters is kept in the experimental dataset, e.g. only the average wind speed is retained, while the values captured by each of the two nacelle anemometers are removed.

Removing Noise. Considering that we are dealing with a real-world dataset, it is expected that the data will contain a substantial amount of noise, e.g. outliers, extreme values, etc., which will impact negatively the outcome of the mining if they are not removed. Several different filters based on the most important output parameter active power are applied in order to remove points with an unlikely active power based on their input parameters, by considering each wind turbine separately.

In Fig. 2 one can see the effect of this cleaning approach on the power curve based on data from one of the wind turbines in the fleet.

Standardisation. The different parameters monitored have values with very different ranges (e.g. the generator bearing temperature varies between −5 and 80 degrees of Celsius, while the generator speed has values between 0 and 1800 rpm), and are of different nature (angular versus non-angular). This makes it very difficult to compare and estimate similarity between parameter values (feature vectors) in time since most of the distance metrics will not perform well.

---

[1] Supervisory control and data acquisition (SCADA) is an architecture to control industrial systems by use of both external and internal sensors (sources).

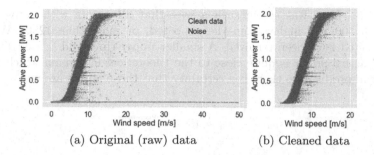

(a) Original (raw) data   (b) Cleaned data

Fig. 2. Power curve of the remained and cleaned points in one of the wind turbines.

Therefore, angular parameters are transformed into two non-angular values by there sine and cosine value. In the case of the wind direction parameter, we multiply the sine and cosine values with their wind direction. By doing this the information of both wind speed and wind direction are captured into the two new variables. Additionally, min-max normalisation [9] is applied on the parameters across the time window selected for analysis, per wind turbine. In this way, all parameter values are scaled relatively within the same turbine between 0 and 1, which is resulting in much more homogeneous feature vectors per timestamp.

5.2 Implementation and Availability

The proposed Layered Multi-view Analysis methodology has been implemented in Python version 3.6. In our experiments we have used four different cluster validation measures: Silhouette Index, Calinski-Harabasz Index, Davies-Bouldin Index and Connectivity. The first three indices and k-means clustering are used from the Python library Scikit-learn. Connectivity Index, min-max normalization and hypercube binning algorithm have been implemented in Python according to their original descriptions (see Sect. 5.1). Methods from Python Matplotlib and Seaborn libraries are used for visualisation. We have also used the implementation of KDE and Silverman rule provided by Python SciPy. Finally, Python Pandas library is used for its DataFrame implementation and NumPy library for a couple of mathematical manipulations.

The executable of the Layered Multi-view Analysis algorithm and the experimental results are available on GitHub. The datasets can be found in the website of Engie.

6 Results and Discussion

The original public SCADA data of a fleet of 4 turbines have been downloaded and pre-processed individually per turbine applying the different steps described in Sect. 5.1. The binning method used for the removal of outliers utilised bin widths of 0.33 m/s for wind speed and 9 degrees for wind direction. The extreme active power filter was set in such a way that all points with active power more than 500 kW higher than the expected, or more than 1250 kW lower than expected have been removed. A smaller upper threshold is used since the expected active power is quite close to the theoretical maximum which a wind turbine can produce, so there is a certainty that those points are noise.

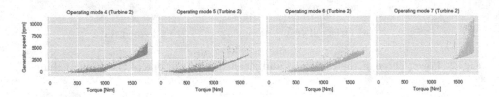

Fig. 3. Torque curves of the operating modes for one of the wind turbines.

In summary, our pre-processed experimental dataset is covering a period of about 8 years and is split in 4 different datasets, one per turbine in the fleet. In this section we represent and discuss the results obtained by applying the proposed multi-view data analysis approach as outlined in Sect. 4.5.

6.1 Individual Analysis Layer: Operating Mode Characterisation

This layer is concerned with the internal working of wind turbine. The following selection of 12 endogeneous parameters, which are the ones retained after eliminating correlated parameters (Sect. 5.1), are considered: sine and cosine of the pitch angel, generator speed, generator bearing temperature 1 and 2, generator stator temperature, gearbox bearing 1 and 2 temperature, gearbox inlet temperature, gearbox oil sump temperature, rotor bearing temperature and torque.

In what follows, we will refer to the 12 parameters as p_1, p_2, ..., p_{12} following the order in which they are listed. Subsequently, the k-means clustering algorithm has been applied on the 4 datasets, one per turbine, composed of the 12 parameters. The optimal amount of clusters (k) per turbine was determined by applying a majority voting (Sect. 4.1), resulting in $k = 3$ for two of the turbines and $k = 4$ for the other two. The difference between the obtained clusters is illustrated in terms of the behaviour of the torque curve (torque as a function of the generator speed), as depicted for one of the four turbines in Fig. 3. The torque curve being derived from an endogeneous parameter is better suited to illustrate difference in operational behaviour rather than the most frequently used power curve.

In total, 14 clusters (operating modes) have been derived. The assumption is that each cluster is representing a distinctive operating mode of the turbine. Each operating mode is characterised in terms of the allowable ranges of each of the 12 internal parameters. Those can be consulted on our GitHub repository.

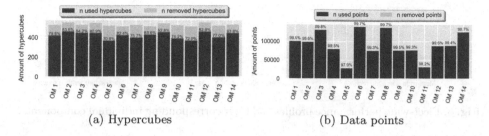

(a) Hypercubes   (b) Data points

Fig. 4. Percentage of retained data per cluster after removal of sparse hypercubes.

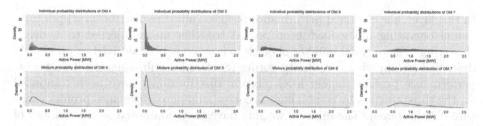

Fig. 5. Performance profiles per operating mode for wind turbine 2.

6.2 Mediation Layer: Performance Profiling

In this layer, performance profiles are derived for each operating mode following the steps outlined in Sect. 4.5. The corresponding values for temperature, wind speed and wind direction (after their non-angular transformation as stated in Sect. 5.1) per cluster are binned together using the hypercube approach (see Sect. 4.3) and the corresponding active power values per hypercube are used to compute a KDE using Gaussian kernel with Silverman's rule (see KDE Sect. 4.2).

Although it was expected that the KDE computation might be influenced by the number of points in each hypercube, or indirectly by the binning granularity, experiments with different sizes of the hypercubes demonstrated very robust KDE computation w.r.t. varying bin sizes. The results presented in the study have been obtained by splitting the solution space into 2250 equal size hypercubes, where each operating mode has around 500 hypercubes containing data points. Subsequently, sparse hypercubes (with less than 10 points) have been removed for the sake of statistical representativeness. The latter did not lead to substantial information loss since as it can be witnessed in Fig. 4, the retained around 5% of the hypercubes for each cluster contain more than 97% of the original data points.

Subsequently, the mixture probability distribution for each cluster (operating mode) is derived as outlined in Sect. 4.5. Figure 5 depicts for one of the turbines the individual probability distributions derived for the different hypercubes and the corresponding mixture probability distributions per cluster.

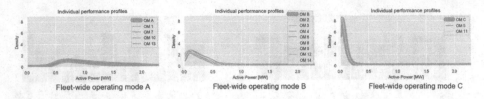

Fig. 6. Fleet-wide performance profiles and their corresponding individual components.

6.3   Integration Layer: Fleet-Wide Performance Labelling

In this layer, the obtained performance profiles (mixture distributions) per individual operating mode have been pooled together and subjected to k-means clustering. The optimal number of clusters $k = 3$ has been derived as previously described by applying a majority voting (Sect. 4.1). Subsequently, 3 fleet-wide performance profiles (higher level mixture distributions) have been computed for the three clusters by combining the corresponding performance profiles (mixture distributions). The mixture weights have been computed as the number of points in the corresponding cluster from layer 1, normalised by the total number of points in the given fleet-wide cluster. The resulting very distinctive fleet-wide performance profiles (A, B and C) are depicted in Fig. 6.

Note that each of the fleet-wide performance profiles can be traced back to a subset of individual operating modes (by use of the table constructed in Sect. 6.1 and available for consultation on our GitHub repository), resulting in fleet-wide (composite) operating modes, which we also denote with A, B and C: $A = \{1, 3, 5, 6, 8, 9, 12, 13\}$; $B = \{2, 4, 10, 14\}$; $C = \{7, 11\}$. It is interesting to observe that the composite operating mode linked to profile C can be traced back to only two of the four wind turbines.

The derived fleet-wide (composite) operating modes, each associated with a very distinctive performance profile (see Fig. 6), can now be used to label the fleet data as follows: 1) for each timestamp, consider the values of the 12 operational parameters; 2) determine to which operating mode they can be assigned (based on the table constructed in Sect. 6.1); 3) identify the composite operating mode to which the identified mode belongs; 4) subsequently, assign the corresponding letter A, B, C or D (not seen) to the timestamp. In this way, each dataset per turbine can be converted into A, B, C or D code (as a DNA sequence), which can be very insightful for monitoring purposes (e.g. long periods of B would signify optimal performance), but is also a powerful representation enabling more advanced applications, e.g.: mining the fleet data for interesting patterns such as transitions between operating modes; zooming in periods with too many Ds; training a predictor of expected production on historical data to be used to detect deviations during real-time operations.

7   Conclusion and Future Work

We have proposed a novel data analysis approach that can be used for multi-view analysis and integration of heterogeneous real-world datasets originating from multiple sources. The validity and the potential of the proposed approach has been demonstrated on a real-world dataset of a fleet of wind turbines. The obtained results are very encouraging. The method is very efficient and robust in detecting characteristic operating modes across the fleet. Subsequently, distinctive performance profiles are derived and associated with each operating mode, which enable converting the fleet data into powerful letter code suitable for more advanced mining.

For future work, we are interested to extend our research in the following directions: 1) fine-tune the method by using e.g. an adaptive hypercube binning; 2) testing different experimental scenarios e.g. comparing different time periods from the same wind turbine; 3) consider additional validation use cases dealing with multi-source datasets e.g. mobility or manufacturing data; 4) extend further the method by exploiting the possibility to covert the fleet data into letter code.

References

1. Bickel, S., Scheffer, T.: Multi-view clustering. In: ICDMM, vol. 4, pp. 19–26 (2004)
2. Boeva, V., Tsiporkova, E., Kostadinova, E.: Analysis of multiple DNA microarray datasets. In: Kasabov, N. (ed.) Springer Handbook of Bio-/Neuroinformatics, pp. 223–234. Springer, Heidelberg (2014). https://doi.org/10.1007/978-3-642 30574-0_14
3. Caliński, T., Harabasz, J.: A dendrite method for cluster analysis. Commun. Stat. Theory Methods 3(1), 1–27 (1974)
4. Davies, D.L., Bouldin, D.W.: A cluster separation measure. IEEE Trans. Pattern Anal. Mach. Intell. 2, 224–227 (1979)
5. Deepak, P., Jurek-Loughrey, A. (eds.): Linking and Mining Heterogeneous and Multi-view Data. USL. Springer, Cham (2019). https://doi.org/10.1007/978-3-030-01872-6
6. Dong, X.L., Srivastava, D.: Big data integration. In: 2013 IEEE 29th International Conference on Data Engineering (ICDE), pp. 1245–1248. IEEE (2013)
7. Gamberger, D., Mihelčić, M., Lavrač, N.: Multilayer clustering: a discovery experiment on country level trading data. In: Džeroski, S., Panov, P., Kocev, D., Todorovski, L. (eds.) DS 2014. LNCS (LNAI), vol. 8777, pp. 87–98. Springer, Cham (2014). https://doi.org/10.1007/978-3-319-11812-3_8
8. Handl, J., Knowles, J., Kell, D.B.: Computational cluster validation in postgenomic data analysis. Bioinformatics 21(15), 3201–3212 (2005)
9. Liu, Z., et al.: A method of SVM with normalization in intrusion detection. Procedia Environ. Sci. 11, 256–262 (2011)
10. MacQueen, J., et al.: Some methods for classification and analysis of multivariate observations. In: Proceedings of 5th Berkeley Symposium on Statistics and Probability (1967)
11. Rousseeuw, P.J.: Silhouettes: a graphical aid to the interpretation and validation of cluster analysis. J. Comput. Appl. Math. 20, 53–65 (1987)
12. Sheather, S.J.: Density estimation. Stat. Sci. 19, 588–597 (2004)

13. Silverman, B.W.: Density Estimation for Statistics and Data Analysis, vol. 26. CRC Press, Boca Raton (1986)
14. Xu, C., Tao, D., Xu, C.: A survey on multi-view learning. arXiv preprint arXiv:1304.5634 (2013)
15. Zhang, J.: Multi-source remote sensing data fusion: status and trends. Int. J. Image Data Fusion **1**(1), 5–24 (2010)

Real-Time Outlier Detection in Time Series Data of Water Sensors

L. van de Wiel[1,2](✉) [ID], D. M. van Es[2] [ID], and A. J. Feelders[1] [ID]

[1] Department of Information and Computing Sciences, Utrecht University, Princetonplein 5, 3584 CC Utrecht, The Netherlands
luukvandewiel@hotmail.com
[2] Ynformed, Stadsplateau 4, 3521 AZ Utrecht, The Netherlands

Abstract. Dutch water authorities are responsible for, among others, the management of water levels in waterways. To perform their task properly, it is important that data is of high quality. We compare several univariate and multivariate methods for real time outlier detection in time series data of water sensors from Dutch water authority "Aa en Maas". Their performance is assessed by measuring how well they detect simulated spike, jump and drift outliers. This approach allowed us to uncover the outlier parameter values (i.e. drift or jump magnitude) at which certain detection thresholds are reached. The experiments show that the outliers are best detected by multivariate (as opposed to univariate) models, and that a multi-layer perceptron quantile regression (QR-MLP) model is best able to capture these multivariate relations. In addition to simulated outliers, the QR-MLP model is able to detect real outliers as well. Moreover, specific rules for each outlier category are not needed. In sum, QR-MLP models are well-suited to detect outliers without supervision.

Keywords: Outlier detection · Time series · Quantile regression · Synthetic evaluation · Machine learning

1 Introduction

Data validation is an important issue for water authorities in the Netherlands. These regional government bodies are responsible for, among others, sewage treatment, dyke management and the management of water levels in waterways. It has been shown that validation pipelines along with implementation advice result in more reliable policy advice, improved operational management and enhanced assessment of current management practices [18]. We examine water data from *Waterschap Aa en Maas*, one of the 21 water authorities in the Netherlands.

To improve data quality, we try to separate outliers from 'real' data points. Our focus is on real-time outlier detection in time series of water sensor measurements. The sensor data consists of time series with fixed intervals between measurements. Different sensors can output time series that are correlated with each other. Here,

Made possible by Ynformed and Waterschap Aa en Maas.

V. Lemaire et al. (Eds.): AALTD 2020, LNAI 12588, pp. 155–170, 2020.
https://doi.org/10.1007/978-3-030-65742-0_11

we can use time series from one or multiple sensors to predict other sensor values. If a big difference between the predicted and observed value occurs, the value may be classified as an outlier [1]. It is important that outliers are detected in real-time, as it enables taking immediate action to resolve possible issues, such as misbehaving sensors or a change in the sensor environment.

Our research focuses on finding which methods can be applied to detect outliers in an unlabelled, unvalidated data set of multivariate time series in a real-time setting. The data is unvalidated; it is raw sensor data that has not gone through any processing steps to improve quality. The data is generally also unlabelled, which means that domain experts have not indicated whether outliers occur. An exception to this is in a few time series that we used for analysis.

We compare different regression-based methods, that predict sensor values given (1) only the sensors history ('univariate'), or (2) given only measurements of other sensors ('multivariate'). Outliers are then determined when the observed data deviates too much from the predicted value. The univariate approach is simpler and can be implemented more easily in practice. Yet, this method runs the risk of carrying past outliers (such as drift) into the future. This would then correctly predict outlying sensor behaviour, thereby failing to label it as outlying. We expect the multivariate approach to solve this problem, as it is not informed about the potentially outlying target history.

2   Data Overview

The data from Aa en Maas comprises water height data in weirs, with a measurement frequency of 15 min. At these weirs, we have access to water height on the upper part and the lower part of the weir, and also to water flow rate and weir shutter height. We used water heights on the *upper* part of the weirs for the analysis. These time series (which are the exceptions described in Sect. 1) were designated by domain experts as not containing any outliers. We gathered all data between 05-06-2015 and 01-07-2019.

An example of water height data on the upper part of a specific set of weirs is shown in Fig. 1[1]. We see that this data is not without errors. For example, the "108HOL_upper" time series (bottom line) has a strange swing around October 2018. Some other minor spikes can also be encountered in this same series. Furthermore, missing values can occur, as seen near the end of the "108IJZ_upper" time series. This data set has relatively few missing values (approximately 2500); other sensor sets have more.

The data of Fig. 1 is of relatively high quality. However, when looking at other sensor sets, time series seem more noisy. In addition, other sensors had more missing data. To evaluate outlier detection capabilities under varying data quality conditions, we selected multiple sensors that had varying data qualities. We chose four weirs for the analysis and model evaluation. These weirs are 102BFS, 103HOE, 104OYE and 201D. For each target time series, we used four other time series as features in multivariate modelling, see Sect. 3.1.

---

[1] This data is not used in our main experiments.

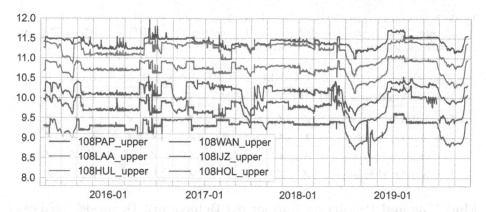

Fig. 1. An example of available water height data. These six sensors are all on the same body of water and are relatively close to each other (about 6 km as biggest distance).

3 Experiment Setup

3.1 Outlier Detection Pipeline

Sensor Selection. We first discarded all sensors that had more than 10% missing values. Then, for each target sensor, from the hundreds of sensors we selected four sensors that correlated most with the target and used those as predictors. This ensures decent model performance while reducing the danger of overfitting.

Imputation. The data set contained missing values, which can be the result of sensor network issues or sensor malfunctioning, for example. We imputed rather than discarded these values to ensure evenly spaced time series. To determine the best suited imputation method for our problem, we benchmarked various methods. For this, we simulated gaps (of similar duration distributions compared to the actual missing gaps) in the time series and measured how well a MICE procedure [17] with different estimators (extra trees, linear regression, Bayesian ridge, KNN, random forests and MTSDI [8]) was able to reproduce the missing values. This showed that the linear regression estimator worked best.

Feature Engineering. For most multivariate experiments, we used rolling lag, min, max and mean features with time steps of 15 and 30 min and 1, 2, 4, 8 and 16 h. In the univariate setting, only the mean values over a prolonged period of time turned out to be useful. For most univariate models we used the mean values of window sizes [64 h, 128 h, ..., 1048 h]. The used features per model type are described in Table 1.

Feature Scaling. To stabilise and enhance model training, we scaled all features to unit variance and zero mean.

Table 1. Features per model type. Models are described in Sect. 4. The five multivariate models are linear regression (LR), MLP, Perceptron (P), QRF and RNN. The five univariate models are LR, MLP, P, AR and isolation forests (IF).

Algorithm	LR	MLP	P	QRF	RNN	LR	MLP	P	AR	IF
Uni-(U)/Multivariate(M)	M	M	M	M	M	U	U	U	U	U
Use feature engineering	✓	✓	✓	✓	✗	✓	✓	✓	✗	✓
Use raw lag values	✗	✗	✗	✗	✓	✗	✗	✗	✓	✗
Use target sensor itself	✗	✗	✗	✗	✗	✓	✓	✓	✓	✓
Use correlated sensors	✓	✓	✓	✓	✓	✗	✗	✗	✗	✗

Modelling and Predicting (ab)normal Behaviour. The models used can be divided into two categories. The first one is regression-based models, where a prediction for a target variable is made (possibly accompanied by quantile values). We can compare this against the actual value and then calculate residuals.

The other category is direct classification. This approach looks at data and then directly determines whether it is an outlier or not.

Outlier Classification. Most of our regression-based models use quantile regression. To perform outlier detection when using these models, we applied the Western Electric rules [19]. We applied Rule 1 and a variation of Rule 2. Rule 1 indicates a single point that falls outside of the 3σ-limit as outlying. Rule 2 does this if two out of three successive points fall beyond the 2σ-limit.

The original Rule 2 led to a high number of false positives. Our improved approach was to look at predictions averaged over the span of a day, and check whether this exceeds the averaged values of the 2nd quantile. Minor short-lived errors now get smoothed out and we get a more accurate way of describing a gradual change. This is described in Algorithm 1. The first three lines downsample the target time series and the upper and lower 2nd quantile (which were outputted by the quantile regression model) from a frequency of 15 min to daily data. The next line performs the detection: if the downsampled time series is above the upper limit, or below the lower limit, an outlier is classified. Eventually, this data is upsampled to a frequency of 15 min and returned.

Algorithm 1. Drift detection by downsampling.

1: **function** DRIFT_DETECTION($y_{in}, q2_{upper}, q2_{lower}$)
2:     $y_{daily} \leftarrow$ DOWNSAMPLE_TO_DAY(y_{in})
3:     $q2_{upper} \leftarrow$ DOWNSAMPLE_TO_DAY($q2_{upper}$)
4:     $q2_{lower} \leftarrow$ DOWNSAMPLE_TO_DAY($q2_{lower}$)
5:     $outliers\_daily \leftarrow (y_{daily} > q2_{upper}) \cup (y_{daily} < q2_{lower})$
6:     **return** UPSAMPLE_TO_15MIN($outliers\_daily$)   ▷ Outliers per 15 minutes.

3.2 Synthetic Evaluation

Correctly recognising outliers is crucial. Therefore, we decided to use the F_β-score with $\beta = 2$ to evaluate performance.

Experts of the water authority established that no outliers are present in the test data of the four time series. This is beneficial for the synthetic evaluation, as already present outliers might interfere with the ones we introduce. The synthetic evaluation method entails that we altered the data to simulate outliers that might happen in reality. Such a method has been applied before in the literature [14].

We studied common outlier definitions to get an idea for outlier categories in water time series data [10,18]. We focused on three synthetic outlier categories because they were regarded to be important by the domain experts:

- **Jumps:** A period of data which is increased or decreased by a constant value. After the period has ended, the data values return back to the original range.
- **Extreme values:** Isolated data points which are increased or decreased by a constant value.
- **Linear drift:** The occurrence of a series which has a gradual linear trend upwards or downwards.

To perform synthetic evaluation for jumps and linear drift, we created multiple test series with different outliers in it. We used one specific drift or jump and then moved this outlier throughout the data, with each movement yielding a new series. We alternated between outliers oriented upwards and downwards. For each test case, we created 100 of these series. We used multiple outlier generation values (in meters), which were 0.02, 0.05, 0.1, 0.2 and 0.3. Jumps had the duration of approximately 1.5 months, whereas drifts lasted for approximately 6 months. Examples are shown in Fig. 2.

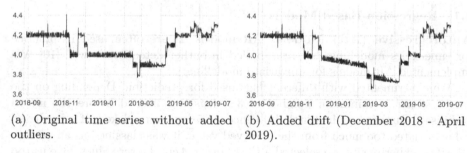

(a) Original time series without added outliers.

(b) Added drift (December 2018 - April 2019).

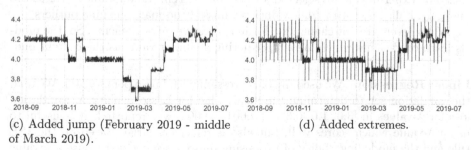

(c) Added jump (February 2019 - middle of March 2019).

(d) Added extremes.

Fig. 2. Outlier examples in sensor 104OYE for outlier value 0.2. For jump and drift, this is 1 of the 100 created series.

4   Modelling and Hyper-Parameter Tuning

In this section we give a short description of the models and algorithms used in this study, and how we tuned their hyper-parameter settings. Input features were used as described in Table 1. The data set is divided into training (60% of the data), validation (20%) and testing (20%) sets. The training data is used to fit different models which are described in this section. The validation data is used to perform hyper-parameter tuning. The test data is used to assess model performance. Model evaluation is performed through a synthetic outlier approach (see Sect. 3.2). According to the 60%–20%–20% split, this means that the training data is in the range of 05-06-2015–13-11-2017, the validation data ends at 06-09-2018 and the testing data ends at 01-07-2019.

An important distinction is between univariate models that only use a sensor's own history to predict future values, and multivariate models that use the values of other sensors. An advantage of the univariate approach is that it is always applicable as no other time series are needed. Furthermore, large sudden changes in values might be easy to track. If we use a multivariate approach, we ignore a sensor's own history and base the detection on other sensor time series. We do this to prevent consistently predicting the same value as currently present (working like a persistence model), which will fail to detect drift. An advantage of the multivariate approach is that we can detect (gradual) changes which happen in only one sensor. If a sensor is slowly drifting, for example, a multivariate approach could detect this based on data from other sensors, whereas a univariate approach may be unable to detect this successfully.

4.1   Regression-Based Models

Autoregressive (AR) Models. AR models [2] are often used in practice for time-series modelling, but are applied in outlier detection as well [10]. We implemented AR models for univariate modelling.

We experimented with different lags used for prediction. Depending on the target sensor, a minimum number of 3–5 lags was needed before the model stabilised. Further lags had little influence, so 5 lags were picked. If a predicted value deviated too much from the observed value, it was classified as an outlier. Based on experiments, we selected a threshold of 4 cm. Lower values gave us too many false alarms, with higher values we missed too many genuine outliers.

Water levels are not changing a lot in successive measurements. So, the models learn coefficients which favour predicting a similar value as the current one.

Linear Regression. We used linear regression with Lasso penalty [16]. We used the validation set to determine the ideal value for Lasso penalty λ. Validation loss for λ-values in [3.0, 1.0, 0.3, ..., 0.001, 0.0003] was reported. It is useful to use a λ-value which scores well, but also is relatively large. Based on the 1SE rule and the modelling ability of promising sensors, we decided to use a λ-value of 0.03 throughout the experiments.

An issue when using linear regression for outlier detection is in defining the outlier detection threshold. We opted for a quantile approach so we could use the same classification rules as in Sect. 3.1. To calculate the quantiles, we added or subtracted the standard deviation of the target time series multiplied by a scalar value to the prediction. This is shown in Eq. 1.

$$q_i = \hat{y} \pm \frac{i}{2}\hat{\sigma}(y_{train}) \quad | \quad i \in \{1, 2, 3\} \tag{1}$$

A disadvantage is that this will lead to a fixed quantile width for the whole model. Varying quantile width is desirable, as uncertainty about the predictions can differ throughout the data.

Quantile Regression Forests (QRF). Parameter values of the QRF algorithm [13] were based on experiments. We used 1000 different trees in total. For each tree, we used the same parameter settings: A node needs to have at least 40 samples in it for it to be considered for a split, a resulting leaf node must have at least 20 samples and the maximum number of considered features per split is one third of the total number of features.

4.2 Neural Network-Based Approaches

To perform neural network architecture tuning systematically, we used the Hyperband algorithm [11]. In the multivariate experiments we averaged the predictions of 10 different neural networks. This ensemble approach is chosen as random weight initialisation has a sizeable effect on the model. In the univariate experiments the ensemble size is lowered to 5, to keep running times acceptable. Many extra predictions are needed because the input of the testing data changes for each outlier time series, which was not the case in the multivariate modelling.

Quantile Regression: Multi Layer Perceptron. The quantile regression multi layer perceptron (QR-MLP) model is a neural network with hidden layers that only uses dense layers. Multiple output nodes are used to calculate values for different quantiles. We use the pinball loss function [9,15] where all the quantiles are taken into account. In the algorithm runs, we used early stopping with a patience value of 5 and a mini-batch size of 128.

The Hyperband algorithm used 5 executions per trial, 3 Hyperband iterations, a factor of 3 and max epochs of 30. In the end, 270 trials were run. It selected the number of layers (1, 2, 4 or 8), number of neurons per layer (16, 32, 64, 128 or 256), dropout (0.0, 0.1, 0.2, 0.3 or 0.4) and learning rate of the network (0.005, 0.001, 0.0005, 0.0001, 0.00005 or 0.00001).

It was not possible to find one general network architecture that works in all cases. There seems to be some correlation between the validation loss and the network complexity. For example, sensor 104OYE can be modelled relatively well and only uses one layer. On the other end of the spectrum we see 102BFS (which was selected to test the impact of its low correlation with other sensors), which has high validation loss and needs more complex models. We decided to use a different architecture for each sensor. Due to computational and time restraints, we were not able to optimise different numbers of neurons per layer. The resulting architectures are shown in Table 2.

Table 2. QR-MLP model architectures per sensor.

Sensor	Dropout	Learning rate	Number of layers	Neurons per layer	Average validation loss of final model
104OYE	0.4	0.0005	1	128	0.1820
103HOE	0.0	0.005	1	256	0.8790
201D	0.4	0.005	2	128	0.9580
102BFS	0.4	0.00005	8	128	1.5330

Quantile Regression: Perceptron Model. A baseline neural network model in the form of a QR-perceptron model was created. This network has no hidden layers. It is somewhat similar to the linear regression model, but like in the QR-MLP model, we use the pinball loss function with multiple output nodes. We thus still have varying quantile width. The only hyper-parameter that needs to be tuned is the learning rate. An exhaustive grid search is now possible. Experiments showed that a relatively large learning rate of 0.005 works best for this kind of model. This value was used for all the QR-perceptron models.

Quantile Regression: RNNs. These networks used RNN layers instead of dense layers. We let the tuner decide if a GRU [3] or LSTM [7] kind of RNN layer should be used. For speed, we now use at most 4 layers, a batch size of 2048 and a window size of 32. A difference with the other multivariate approaches, is that since we have a RNN, all these 32 values are used in every step. Also, this disallows us from explicitly modelling features like the minimum and mean features.

Again, there did not seem to be a best overall network architecture. Moreover, it seems that neither LSTM- nor GRU-layers work best for every network. We use a different network architecture per sensor, as shown in Table 3.

Table 3. RNN model architectures per sensor

Sensor	RNN type	Dropout	Learning rate	Number of layers	Neurons per layer	Average validation loss of final model
104OYE	GRU	0.1	0.0005	1	256	0.2060
103HOE	LSTM	0.0	0.005	1	256	0.9401
201D	LSTM	0.4	0.001	4	64	1.3292
102BFS	LSTM	0.2	0.0001	4	64	1.2330

4.3   Direct Classification Model: Isolation Forests (IF)

The isolation forest model [12] is often applied in the literature [5]. When using multivariate feature sets, we can only look at outliers of a whole system (like a group of 5 sensors), instead of at outliers of a single sensor. Also, it is mandatory to incorporate the history of the target sensor. Since we want to know if a specific sensor is behaving strangely, this method is only suited for our univariate setting.

We have performed hyper-parameter tuning to determine the ideal value of the contamination parameter. If we set the contamination value too low, we will detect few outliers. If it is set too high, the precision of our model will drop. Our experiments suggested a value of 0.07.

5   Results

In this section, we first describe illustrative examples for the univariate and multivariate models. Then, we compare these models. We end this section with a description of the practical impact of the best performing ones.

5.1   Illustrative Examples: Univariate Results

The QR-RNN and QRF models have not been applied to the univariate modelling experiments, since this became prohibitively slow. We applied the linear regression, QR-MLP, QR-Perceptron, AR and IF models here. We show the visual results of one specific time series of sensor 104OYE in Fig. 3. We added a jump of 0.2 m from February 2019 to the middle of March 2019.

Results QR-MLP. For the univariate QR-MLP models, we used the same architecture for each sensor, as we do not have to take into account correlated time series. This was the same architecture that was used for 104OYE in multivariate QR-MLP modelling, as described in Table 2. Figure 3a shows that some parts of the added jump can be detected, but this is certainly not the case for the sequence of outliers as a whole.

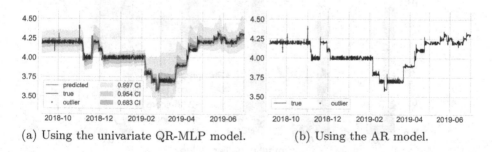

(a) Using the univariate QR-MLP model.   (b) Using the AR model.

Fig. 3. Plots of sensor 104OYE, with added jump.

Results AR. AR model performance is shown in Fig. 3b. We see here that the begin and end points of the added outlier sequence can be detected. The period in between can not be detected, though. Also, some other sudden changes in the time series have been classified as outliers.

5.2 Illustrative Example: Multivariate Results QR-MLP

We compared the multivariate linear regression, QR-MLP, QR-Perceptron, QRF and QR-RNN models. Figure 4 shows the results of the same 104OYE time series with added jump, now modelled multivariately by QR-MLP. This jump is detected well, but some false positives are also present. Some outliers may have been missed by the domain experts. This is most visible around November 2018. Results of drifts detection for all outlier values are shown in Fig. 5.

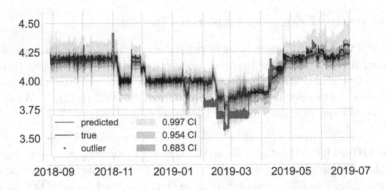

Fig. 4. Quantile plot of sensor 104OYE, with added jump using the QR-MLP model.

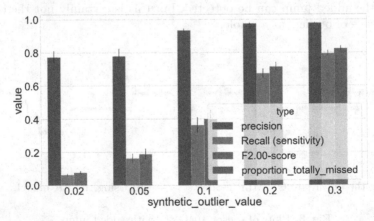

Fig. 5. Bar plots of QR-MLP model scores of sensor 104OYE for all outliers values for drifts. The legend shows the *proportion totally missed*, which indicates the proportion of the outlier sequences missed completely. This value is 0 for each outlier value.

5.3 Comparison of Univariate and Multivariate Modelling Techniques

The F_2 scores of all models are shown in Fig. 6. This figure shows how well different multivariate and univariate models score on different outlier categories for different sensors. We see some clear differences. To compare the performance of these different models, we followed a two-step approach [4]. First, a Friedman Aligned Ranks test was performed to check whether there are significant differences between the distributions of the results of the models [6]. If this test yielded a significant difference, we used the Nemenyi test to compare all models pairwise.

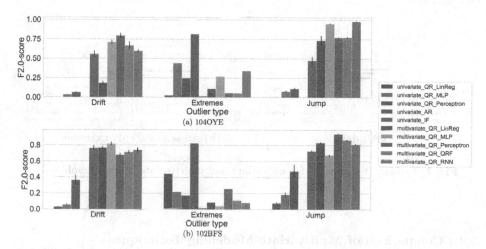

Fig. 6. Bar plots of model performance on all data sets, univariate and multivariate combined for outlier value 0.2.

We used the Friedman Aligned Ranks test in 8 scenarios. One of them consisted of all sensors and outlier categories and sizes. We also divided the data by outlier category and by sensor. All these distributions are significantly different from each other (see Table 4). This is expected, as we compare many models which have very different performances as is visible in Fig. 6.

Table 4. Friedman Aligned Ranks results ($\alpha = 0.05$) of all experiments.

Sensors	All	All	All	All	102BFS	103HOE	104OYE	201D
Outliers	All	Drift	Jump	Extremes	All	All	All	All
χ^2	118.936	100.000	115.084	60.752	38.274	36.949	39.365	51.748
p-value	0.000	0.000	0.000	0.000	0.000	0.000	0.000	0.000
Significant?	Yes	Yes	Yes	Yes	Yes	Yes	Yes	Yes

We now perform the Nemenyi test to see which models have significantly different performance. The results are shown in Fig. 7. On the horizontal axis, the average ranking of the algorithms is shown. The further a model is to the left on the x-axis, the better it is scoring on average. Algorithms that are connected by a bold line are not differing significantly from each other. In the overall comparison of Fig. 7a, the five multivariate models outperform the five univariate ones. This is because they score better in drift and jump detection scenarios. If we want a single model to detect all outlier types, then the multivariate QR-MLP or QR-perceptron model seems the best choice. In the extreme outlier category (Fig. 7b), however, AR seems to perform exceptionally well. This is due to the fact that AR almost works like a persistence model and can detect a large sudden change easily.

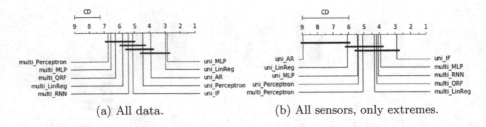

(a) All data.   (b) All sensors, only extremes.

Fig. 7. Nemenyi test results for univariate and multivariate models combined.

5.4   Comparison of Multivariate Modelling Techniques

As the multivariate models outperform the univariate ones in most cases, we zoom in further on the multivariate ones. Figure 8 gives an overview of the F_2-score results of multivariate models for outlier value 0.2 m. Note that the F_2-score for extremes is low in all cases. This may be explained by the fact that fewer outliers are added here than in the other categories. In the drift category, 17520 outlier points are added. In the jump category this number is 4320 and when using extremes, only 100. The total number of true positives differs greatly per method, so the roughly constant number of false positives can severely impact precision and thus the F_2-score.

We see some big differences between the models, but we also note that model performance differs greatly per sensor. Results of the Friedman Aligned Ranks test are shown in Table 5.

In Fig. 9, the Nemenyi test result for all data is shown. QR-MLP and QR-perceptron are significantly different from QR-RNN. Other comparisons showed similar results. Since QR-MLP and the QR-perceptron model perform decently most of the time, these could be go-to algorithms.

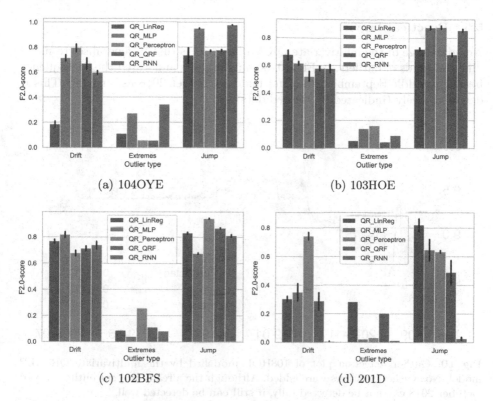

Fig. 8. Bar plots of model performance of multivariate models on all data sets for outlier value 0.2.

Table 5. Friedman Aligned Ranks test results ($\alpha = 0.05$) in multivariate experiments.

Sensors	All	All	All	All	102BFS	103HOE	104OYE	201D
Outliers	All	Drift	Jump	Extremes	All	All	All	All
χ^2	14.947	10.556	8.399	4.384	3.287	4.273	11.549	22.646
p-value	0.005	0.032	0.078	0.357	0.511	0.370	0.021	0.000
Significant?	Yes	Yes	No	No	No	No	Yes	Yes

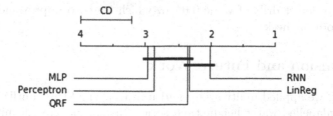

Fig. 9. Nemenyi test results for multivariate models.

5.5 Practical Impact

An advantage of the multivariate QR-MLP model is that it generalises to many different kinds of outliers. In Fig. 10, domain experts annotated the subsequence between middle September 2018 and middle October 2018 as outlying. This is detected nicely (indicated by the red dots).

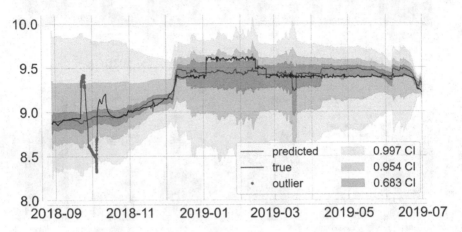

Fig. 10. Outlier detection plot of 108HOL modelled by the multivariate QR-MLP model. No synthetic outliers were added. Although the already present outlier around October 2018 can not be detected fully, it still can be detected well.

Domain experts stated that jump values and extreme values of 0.2 m are reasonable in real life. We can derive from Fig. 6 that the performance of the best performing models is satisfactory in many cases. The domain experts also stated that a drift is generally in between 0.05 m–0.10 m over the period of a year. This roughly corresponds to the two categories of outlier values 0.02 m and 0.05 m. Outliers for these values are harder to detect, as seen in Fig. 5. However, this judgement is (too) harsh, as the algorithms do detect almost every drift sequence after some time. The experts stated that periodical checks for the occurrence of drift are normally performed yearly. Our models need one month on average to detect drift of value 0.05 m, which leads to improvements over a manual periodical check.

6   Conclusion and Future Work

In this work, we applied multivariate and univariate real-time outlier detection models in unlabelled water height time series. Instead of only cleaning historical data, the trained models can be used to monitor sensor measurements and directly signal outlying values. The key contribution of this work is the systematic comparison of algorithms and the easily parametrisable synthetic validation scenarios which were constructed in cooperation with domain experts.

We showed that multivariate approaches work better than univariate approaches for jump and drift outlier types. For extreme values however, univariate approaches appear to outperform multivariate ones. Yet, we think that this result will not hold in practice, because it partly is an artefact of our synthetic evaluation procedure. Univariate outlier models basically function by signalling large instantaneous changes. This indeed highlights extreme values, but may fail to detect slightly more gradual ones. Also, natural (more gradual) jumps in the data will be missed. As we only added instantaneous extreme outliers, many true positives were present. Few other already present data points changed so quickly, so few false positives were present. Thus, this category of models performed well in our simulations. In real life however, extreme values occur less frequently, and natural jumps are more apparent. Therefore, these models are likely to result in inadequate performance when implemented in practice.

It should be noted that the multivariate modelling approach is not applicable for all sensors (like 102BFS). A multivariate approach is only suitable when sufficiently correlating series are available.

Within the category of multivariate models, we found that the QRF approach and the QR-RNN models performed poorly. The QRF model resulted in very jagged quantile boundaries, which resulted in the misclassification of many data points. The QR-RNN model often resulted in very wide quantiles, which worsened performance. Linear regression performed relatively decently, although the fixed quantile width remains an issue. In the end, we can conclude that the QR-MLP and the QR-perceptron models performed the best overall.

To get a better overview how well these different models work in practice, it is recommended that a pilot program is carried out to test the performance in a more practical setting.

Our extreme values scenario had some artefacts. Although we selected a realistic outlier value in cooperation with the domain experts, it is worthwhile to investigate more realistic scenarios. An example is a more gradual extreme value. This is fundamentally different from drift, as a gradual extreme value could occur in a few time steps, in contrast to a duration of multiple months. Research into a combination of different (extreme) outlier categories may also be useful.

It may be a fruitful idea to use different models to detect different outlier categories. For example, combining the results of an AR model and a multivariate QR-MLP model could work to detect extreme values, jumps, and drifts.

An interesting research subtopic regards determining outlier causes. An outlier can be caused by multiple factors. Different kinds of outliers might require different means of alleviation. It is of interest to determine these different causes with additional techniques.

Another noteworthy subtopic concerns propagating sensor errors. If a sensor malfunctions, this will not only affect its own predictions, but will affect all other sensor predictions that make use of the values of this malfunctioning sensor as a predictor variable as well. Further research is needed to make accurate claims about this phenomenon.

References

1. Aggarwal, C.C.: Data Mining: The Textbook. Springer, New York (2015). https://doi.org/10.1007/978-3-319-14142-8
2. Chatfield, C.: The Analysis of Time Series: An Introduction. Chapman and Hall/CRC, London (2003)
3. Cho, K., et al.: Learning phrase representations using RNN encoder-decoder for statistical machine translation. arXiv preprint arXiv:1406.1078 (2014)
4. Demšar, J.: Statistical comparisons of classifiers over multiple data sets. J. Mach. Learn. Res. **7**, 1–30 (2006)
5. Ding, Z., Fei, M.: An anomaly detection approach based on isolation forest algorithm for streaming data using sliding window. IFAC Proc. Vol. **46**(20), 12–17 (2013)
6. García, S., Fernández, A., Luengo, J., Herrera, F.: Advanced nonparametric tests for multiple comparisons in the design of experiments in computational intelligence and data mining: experimental analysis of power. Inf. Sci. **180**(10), 2044–2064 (2010)
7. Hochreiter, S., Schmidhuber, J.: Long short-term memory. Neural Comput. **9**(8), 1735–1780 (1997)
8. Junger, W., de Leon, A.P.: mtsdi: Multivariate time series data imputation (2012). https://cran.r-project.org/web/packages/mtsdi/index.html. R package version 0.3.5
9. Koenker, R., Hallock, K.F.: Quantile regression. J. Econ. Perspect. **15**(4), 143–156 (2001)
10. Leigh, C., et al.: A framework for automated anomaly detection in high frequency water-quality data from in situ sensors. Sci. Total. Environ. **664**, 885–898 (2019)
11. Li, L., Jamieson, K., DeSalvo, G., Rostamizadeh, A., Talwalkar, A.: Hyperband: a novel bandit-based approach to hyperparameter optimization. J. Mach. Learn. Res. **18**(1), 6765–6816 (2017)
12. Liu, F.T., Ting, K.M., Zhou, Z.H.: Isolation forest. In: 2008 Eighth IEEE International Conference on Data Mining, pp. 413–422. IEEE (2008)
13. Meinshausen, N.: Quantile regression forests. J. Mach. Learn. Res. **7**, 983–999 (2006)
14. Perelman, L., Arad, J., Housh, M., Ostfeld, A.: Event detection in water distribution systems from multivariate water quality time series. Environ. Sci. Technol. **46**(15), 8212–8219 (2012)
15. Rodrigues, F., Pereira, F.C.: Beyond expectation: deep joint mean and quantile regression for spatio-temporal problems. arXiv preprint arXiv:1808.08798 (2018)
16. Tibshirani, R.: Regression shrinkage and selection via the lasso. J. R. Stat. Soc. Ser. B (Methodol.) **58**(1), 267–288 (1996)
17. Van Buuren, S., Groothuis-Oudshoorn, K.: MICE: multivariate imputation by chained equations in R. J. Stat. Softw. **45**(3), 1–68 (2011)
18. Versteeg, R., de Graaff, B.: Valldidatieplan Waterkwantiteitsmetingen. STOWA 2009–20 (2009). (in Dutch)
19. Western Electric Company: Statistical Quality Control Handbook. Western Electric Company, Rossville (1956)

Lightweight Temporal Self-attention for Classifying Satellite Images Time Series

Vivien Sainte Fare Garnot[✉] and Loic Landrieu

LASTIG, ENSG, IGN, Univ Gustave Eiffel, 94160 Saint-Mande, France
vivien.sainte-fare-garnot@ign.fr
https://www.umr-lastig.fr/

Abstract. The increasing accessibility and precision of Earth observation satellite data offers considerable opportunities for industrial and state actors alike. This calls however for efficient methods able to process time-series on a global scale. Building on recent work employing multi-headed self-attention mechanisms to classify remote sensing time sequences, we propose a modification of the Temporal Attention Encoder of Garnot et al. [5]. In our network, the channels of the temporal inputs are distributed among several compact attention heads operating in parallel. Each head extracts highly-specialized temporal features which are in turn concatenated into a single representation. Our approach outperforms other state-of-the-art time series classification algorithms on an open-access satellite image dataset, while using significantly fewer parameters and with a reduced computational complexity.

Keywords: Time sequence · Self-attention · Multi-headed attention · Sentinel satellite

1 Introduction

Time series of remote sensing data, such as satellites images taken at regular intervals, provide a wealth of useful information for Earth monitoring. However, they are also typically very large, and their analysis is resource-intensive. For example, the Sentinel satellites gather over 25 Tb of data every year in the EU. While exploiting the spatial structure of the data poses a challenge on its own, we focus in this paper on the efficient extraction of discriminative temporal features from sequences of spatial descriptors.

Among the many possible approaches to handling time-series of remote sensing data, one can concatenate observations in the temporal dimension [7], use temporal statistics [8], histograms [1], time-kernels [12], or shapelets [16]. Probabilistic graphical models such as Conditional Random Fields can also be used to exploit the temporal structure of the data [2].

Deep learning-based methods are particularly well-suited for dealing with the large amount of data collected by satellite sensors. Neural networks can either

© Springer Nature Switzerland AG 2020
V. Lemaire et al. (Eds.): AALTD 2020, LNAI 12588, pp. 171–181, 2020.
https://doi.org/10.1007/978-3-030-65742-0_12

model the temporal dimension independently of the spatial dimensions with recurrent Neural Networks [4] or one-dimensional convolutions [9], or jointly with convolutional recurrent networks [10] or 3D convolutions [6].

More recently, the self-attention mechanism introduced by Vaswani *et al.* [13], initially developed for Natural Language Processing (NLP), has been successfully used and adapted to remote sensing tasks [5,11]. In Sect. 2.1, we present these approaches and their differences in greater details.

In this paper, we introduce the Lightweight Temporal Attention Encoder (L-TAE), a novel attention-based network focusing on memory and computational efficiency. Our approach is based on the Temporal Attention Encoder (TAE) of Garnot *et al.* [5], with several modifications meant to avoid redundant computations and parameters, while retaining a high degree of expressiveness. We evaluate the performance of our approach on the open-access dataset Sentinel2-Agri [5], consisting of satellite image time series annotated at parcel level (Fig. 1). With an equal parameter count, our algorithm outperforms all state-of-the-art competing methods in terms of precision and computational efficiency. Our method allows for efficient parameters usage, as our L-TAE outperforms TAEs with close to 10 times the parameter count, as well as recurrent units over 300 times larger.

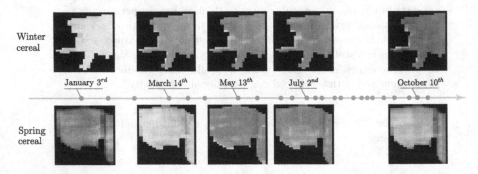

Fig. 1. Example image time series of the Sentinel2-Agri dataset for two parcels of the *Winter cereal* and *Spring cereal* classes, taken from [5]. The dots on the horizontal axis represent the unevenly distributed acquisition dates over the period of interest.

2   Method

Throughout this section, we consider a generic input time series of length T comprised of E-dimensional feature vectors $\mathbf{e} = [e^{(1)}, \cdots, e^{(T)}] \in \mathbb{R}^{E \times T}$. For example, such vectors can be a sequence of learned embeddings of super-spectral satellite images.

2.1   Multi-headed Self-attention

In its original iteration [13], self-attention—initially designed for text translation—consists of the following steps:

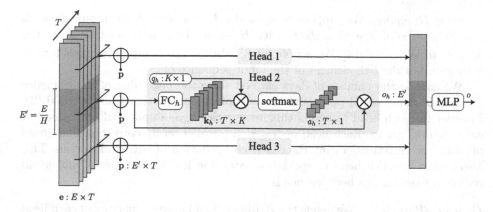

Fig. 2. The proposed L-TAE module processing an input sequence **e** of T vectors of size E, with $H = 3$ heads and keys of size K. The channels of the input embeddings are distributed among heads. Each head uses a learnt query $\hat{q}_h$, while a linear layer FC_h maps inputs to keys. The outputs of all heads are concatenated into a vector with the same size as the input embeddings, regardless of the number of heads.

(i) compute a triplet of key-query-value $k^{(t)}, q^{(t)}, v^{(t)}$ for each position t of the input sequence with a shared linear layer applied to $e^{(t)}$,

(ii) compute attention masks representing the compatibility (dot-product) between the queries at each position and the keys corresponding to previous elements in the sequence,

(iii) associate to each position of the sequence an output defined as the sum of the previous values weighted by the corresponding attention mask.

This process is done in parallel for H different sets of independent parameters—or heads—whose outputs are then concatenated. This scheme allows each head to specialize in detecting certain characteristics of the feature vectors.

Rußwurm *et al.* [11] propose to apply this architecture to embed sequences of satellite observations by max-pooling the resulting sequence of outputs in the temporal dimension. Garnot *et al.* [5] introduce the TAE, a modified self-attention scheme. First, they propose to directly use the input embeddings as values ($v^{(t)} = e^{(t)}$), taking advantage of the end-to-end training of the image embedding functions alongside the TAE. Additionally, they define a single master query $\hat{q}$ for each sequence, computed from the temporal average of the queries. This master query is compared to the sequence of keys to produce a single attention mask of dimension T used to weight the temporal mean of values into a single feature vector.

2.2 Lightweight Attention

We build on this effort to adapt multi-headed self-attention to the task of sequence embedding. Our focus is on efficiency, both in terms of parameter count and computational load.

Channel Grouping: We propose to split the E channels of the input elements into H groups of size $E' = E/H$ with H being the number of heads[1], in the manner of Wu *et al.* [14]. We denote by $e_h^{(t)}$ the groups of input channels for the h-th group of the t-th element of the input sequence (1).

We encode the number of days elapsed since the beginning of the sequence into an E'-dimensional positional vector p of characteristic scale $\tau = 1000$ (2). In order for each head to access this information, p is duplicated and added to each channel group. Each head operates in parallel on its corresponding group of channels, thus accelerating the costly computation of keys and queries. This also allows for each head to specialize alongside its channel group, and avoid redundant operations between heads.

Query-as-Parameter: We define the K-dimensional master query q_h of each head h as a model parameter instead of the results of a linear layer. The immediate benefit is a further reduction of the number of parameters, while the lack of flexibility is compensated by the larger number of available heads.

Attention Masks: As a result, only the keys are obtained with a learned linear layer (3), while values are bypassed ($v^{(t)} = e^{(t)}$), and the queries are model parameters. The attention masks $a_h \in [0,1]^T$ of each head h are defined as the scaled *softmax* of the dot-product between the keys and the master query (4). The outputs o_h of each heads are defined as the sum in the temporal dimension of the corresponding inputs weighted by the attention mask a_h (5). Finally, the heads outputs are concatenated into a vector of size E and processed by a multi-layer perceptron MLP to the desired size (6).

In Fig. 2, we represent a schematic representation of our network. The different steps of the L-TAE can also be condensed by the following operations, for $h = 1 \cdots H$ and $t = 1 \cdots T$:

$$e_h^{(t)} = \left[e^{(t)} \left[(h-1)E' + i \right] \right]_{i=1}^{E'} \tag{1}$$

$$p^{(t)} = \left[sin \left(\text{day}(t)/\tau^{\frac{i}{E'}} \right) \right]_{i=1}^{E'} \tag{2}$$

$$k_h^{(t)} = \text{FC}_h(e_h^{(t)} + p^{(t)}) \tag{3}$$

$$a_h = \text{softmax} \left(\frac{1}{\sqrt{K}} \left[q_h \cdot k_h^{(t)} \right]_{t=1}^{T} \right) \tag{4}$$

$$o_h = \sum_{t=1}^{T} a_h[t] \left(e_h^{(t)} + p^{(t)} \right) \tag{5}$$

$$o = \text{MLP}([o_1, \cdots, o_H]) . \tag{6}$$

2.3   Spatio-Temporal Classifier

Our proposed L-TAE temporal encoder is meant to be learned alongside a spatial encoding module and a decoder module in an end-to-end fashion (7). The spatial

---

[1] E and H are typically powers of 2 and $E > H$, ensuring that E' remains integer.

encoder S maps a sequence of raw inputs $X^{(t)}$ to a sequence of learned features $e^{(t)}$, computed independently at each position of the sequence. The temporal module then maps this sequence to a single embedding o. Lastly, the decoder D maps o to a target vector y, such as class logits for classification.

$$\left[X^{(t)}\right]_{t=1}^{T} \xrightarrow{\ S\ } \left[e^{(t)}\right]_{t=1}^{T} \xrightarrow{\ \text{L-TAE}\ } o \xrightarrow{\ D\ } y \,. \tag{7}$$

3 Numerical Experiment

3.1 Dataset

We evaluate our proposed method with the public dataset *Sentinel2-Agri* [5], comprised of 191 703 sequences of 24 superspectral images of agricultural parcels from January to October, as represented in Fig. 1. The acquisitions have a spatial resolution of 10 m per pixel and 10 spectral bands. Each parcel is annotated within a 20 class nomenclature of agricultural crops.

3.2 Metric and Protocol

We evaluate the performance of our algorithm with two metrics measured at parcel-level: Overall Accuracy (OA) and mean Intersection-over-Union (mIoU), averaged over the class set (macro-averaging).

Given that the dataset is unbalanced (4 classes represent 90% of the samples) the mIoU gives a more faithful assessment of the performance.

We propose two evaluation protocols to assess the efficiency of our proposed light-weight temporal attention encoder:

- We assess the performance of our method and several state-of-the-art parcel classification algorithms on the dataset Sentinel2-Agri. In order to perform a fair comparison, we chose configurations corresponding to around 150k parameters for all methods. We report the results in Table 1 alongside the theoretical number of floating point operations (in FLOPs) required for the sequence embedding modules to process a single sequence at inference time.
- We complement this first experiment by comparing the performance of different configurations of sequence embedding algorithms, and plot the performance with respect to the number of parameters. In order to remove the effects of the different spatial encoders, we use the same spatial encoder (a pixel set encoder [5]) in all experiments. We only adapt the last linear layer of the spatial encoder to produce embeddings of the desired dimensions.

3.3 Evaluated Methods

We evaluate the performance of recent algorithms operating on satellite image time series to assess the relative improvement offered by our method.

PSE+TAE. The approach proposed by Garnot *et al.* [5]. They use a Pixel-Set Encoder (PSE) module to encode each image independently, and process the resulting sequence of embeddings with a TAE. The decoder is a 2-layer MLP.

PSE+L-TAE. Our proposed method. We keep the same architecture as the PSE+TAE, and replace the TAE by our L-TAE network.

CNN+GRU. A similar approach [4] to PSE+TAE, with a CNN instead of the PSE and a Gated Recurrent Unit [3] instead of the TAE.

CNN+TempCNN. Another variation, with a two-dimensional CNN to encode the images and a one-dimensional CNN processing the temporal dimension independently. This architecture is based on the work of Pelletier *et al.* [9].

Transformer. Rußwurm *et al.* were the first to introduce self-attention methods to the classification of remote sensing images. In their work[11], the statistics of images are simply averaged over the parcels' pixels, while the resulting sequence is processed by a Transformer network [13]. The output sequence of embeddings is max-pooled along the temporal dimension to produce a single embedding.

ConvLSTM. Rußwurm *et al.* [10] combine the embedding of the spatial and temporal dimensions by using a ConvLSTM network [15]. This work has been adapted to process parcels instead of pixels [5].

Random Forest. We use the temporal concatenation scheme of Bailly *et al.* to train a random forest of 100 trees using the parcel-wise mean and standard deviation of the spectral bands.

3.4 Analysis

In Table 1, we report the performances of competing methods (taken from [5]) and L-TAE, all obtained with a 5-fold cross-validation scheme. Our L-TAE architecture outperforms other methods on this dataset both in overall accuracy and mIoU. While the OA is essentially unchanged compared to the TAE, the increase of 0.8 mIoU points is noteworthy since our model is not only simpler but also less computationally demanding by almost an order of magnitude.

We would like to emphasize that FLOP counts do not necessarily reflect the computational speed of the model in practice. In our non-distributed implementation, the total inference times are dominated by loading times and the spatial embedding module. However, this metric serves to illustrate the simplicity and efficiency of our network.

Furthermore, our network maintains a high precision even with a drastic decrease in the parameter count, as illustrated in Fig. 3. We evaluate the four best performing sequence embedding modules (L-TAE, TAE, GRU, TempCNN) in the previous experiment with different configurations, ranging from $9k$ to $3M$ parameters. These algorithms all operate with the same decoder and spatial module: a PSE and decoder layer totaling 31k parameters. The smallest L-TAE configuration, with only $9k$ parameters, achieves a better mIoU score than a TAE with almost $110k$ parameters, a TempCNN with over $700k$ parameters, and a GRU with $3M$ parameters. See Table 4 in the Appendix for the detailed configurations corresponding to each points.

Table 1. Performance of our model and competing approaches parameterized to all have 150k parameters approximately. MFLOPs is the number of floating points operations (in 10^6 FLOPs) *in the temporal feature extraction module* and for one sequence. This only applies to networks which have a clearly separated temporal module.

	OA	mIoU	MFLOPs
PSE+L-TAE (ours)	**94.3 ± 0.2**	**51.7 ± 0.4**	**0.18**
PSE+TAE [5]	94.2 ± 0.1	50.9 ± 0.8	1.7
CNN+GRU [4]	93.8 ± 0.3	48.1 ± 0.6	3.6
CNN+TempCNN [9]	93.3 ± 0.2	47.5 ± 1.0	0.81
Transformer [11]	92.2 ± 0.3	42.8 ± 1.1	1.1
ConvLSTM [10]	92.5 ± 0.5	42.1 ± 1.2	–
Random Forest [2]	91.6 ± 1.7	32.5 ± 1.4	—

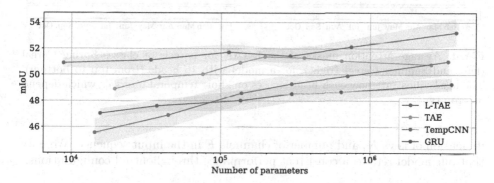

Fig. 3. Performance (in mIoU) of different approaches plotted with respect to the number of parameters in the sequence embedding module. The number of parameters is given on a logarithmic scale. The shaded areas depict the observed standard deviation of mIoU across the five cross-validation folds. The L-TAE outperforms other models across all model sizes, and the smallest 9k-parameter L-TAE instance yields better mIoU than the 100k-parameter TAE model.

In Fig. 4, we represent the average attention masks of a 16-head L-TAE for two different classes. We observe that the masks of the different heads focus on narrow and distinct time-extents, *i.e.* display a high degree of specialization. We also note that the masks are adaptive to the parcels crop types. This suggests that the attention heads are able to cater the learned features to the plant types considered. We argue that our channel grouping strategy, in which each head processes distinct time-stamped features, allows for this specialization and leads to an efficient use of the trainable parameters.

3.5   Ablation Study and Robustness Assessment

In Table 2, we report the performance of our proposed L-TAE architecture with different configurations of the following hyper-parameters: number of heads H,

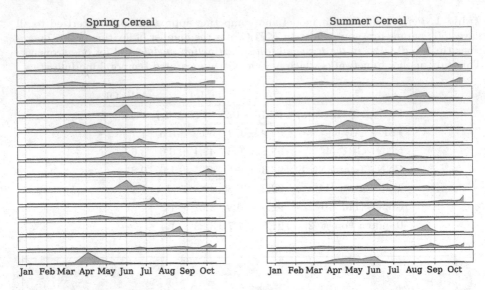

Fig. 4. Average attention masks of the L-TAE for parcels of classes Spring Cereal (left) and Summer Cereal (right), for a model with 16 heads (from top to bottom). The masks illustrate how each head focuses on short temporal intervals which depend on crop type.

dimension of keys K, and number of channels E in the input sequence. We note that our model retains a consistent performance throughout all configurations.

Number of Heads: The number of heads seems to only have a limited effect on the performance. We hypothesize that while a higher number of heads H is beneficial, a smaller group size E' is however detrimental.

Key Dimension: Our experiments show that smaller key dimensions than the typical values used in NLP or for the TAE ($K = 32$) perform better on our problem. Even 2-dimensional keys allow for the L-TAE to achieve performances similar to the TAE.

Input Dimension: The variation in performance observed with larger input embeddings is expected: it corresponds to a richer representation. However, the returns are decreasing on the considered dataset with respect to the number of incurred parameters.

Query-as-Parameter. In order to evaluate the impact of our different design choices, we train a variation of our network with the same master-query scheme than the TAE. The larger resulting linear layer increases the size of the model for a total of 170k parameters, resulting in a mIoU of only 49.7. This indicates that the query-as-parameter scheme is not only beneficial in terms of compactness but also performance.

Table 2. Impact of several hyper-parameters on the performance of our method. Underlined, the default parameters values in this study; in **bold**, the best performance.

H	Params.	mIoU	K	Params.	mIoU	E	Params.	mIoU
2	114k	51.6	2	118k	50.7	32	46k	49.6
4	118k	51.0	4	127k	51.3	64	59k	49.6
8	127k	51.2	<u>8</u>	143k	**51.7**	128	65k	51.1
<u>16</u>	143k	**51.7**	16	176k	50.8	<u>256</u>	143k	**51.7**
32	176k	51.2	32	242k	51.2	512	254k	51.4

3.6 Computational Complexity

In Table 3, we report the asymptotic complexity of different sequence embedding algorithms. For the L-TAE, the channel grouping strategy removes the influence of H in the computation of keys and outputs compared to a TAE or a Transformer. The complexity of the L-TAE is also lower than the GRU's as M, the size of the hidden state, is typically larger than K (130 vs 8 in the experiments presented in Table 1).

Table 3. Asymptotic complexity of different temporal extraction modules for the computation of keys, attention masks, and output vectors. For the GRU, the complexity of the memory update is given in the Keys and Mask columns. X is the size of the output vector. M is the size of the hidden state of the GRU.

Method	Keys	Mask	Output
L-TAE	$O(TEK)$	$O(HTK)$	$O(EX)$
TAE	$O(HTEK)$	$O(HTK)$	$O(HEX)$
Transf	$O(HTEK)$	$O(HT^2K)$	$O(HEX)$
GRU	$O(MT(E+M))$		$O(MX)$

4 Conclusion

We presented a new lightweight network for embedding sequences of observations such as satellite time-series. Thanks to a channel grouping strategy and the definition of the master query as a trainable parameter, our proposed approach is more compact and computationally efficient than other attention-based architectures. Evaluated on an open-access satellite dataset, the L-TAE performs better than state-of-the-art approaches, with significantly fewer parameters and a reduced computational load, opening the way for continent-scale automated analysis of Earth observation.

Our implementation of the L-TAE can be accessed in the open-source repository: github.com/VSainteuf/lightweight-temporal-attention-pytorch.

Acknowledgments. This research was supported by the AI4GEO project: http://www.ai4geo.eu/ and the French Agriculture Paying Agency (ASP).

Appendix

In Table 4, we give the exact configurations used to obtain the values in Fig. 3.

Table 4. Configurations of the L-TAE, TAE, GRU, and TempCNN instances used to obtain Fig. 3.

Parameters	E	H	K	MLP
L-TAE				
9 k	128	8	8	128
34 k	128	16	8	128 - 128
112 k	256	16	8	256 - 128
288 k	512	32	8	512 - 128
740 k	1024	32	8	1024 - 256 - 128
3840 k	2048	64	8	2048 - 1024 - 256 - 128
TAE				
19 k	64	2	8	128 - 128
39 k	64	4	8	256 - 128
76 k	128	4	8	512 - 128
195 k	256	4	8	1024 - 128
360 k	256	4	8	1024 - 256 - 128
641 k	256	8	8	2048 - 256 - 128
2592 k	1024	8	16	8192 - 256 - 128

Parameters	Hidden Size
15k	32
37k	64
134k	156
296k	256
636k	400
3545k	1024
GRU	

Parameters	Kernels	FC
14k	16 - 16 - 16	16 - 16
45k	32 - 32 - 32	32 - 32
136k	64 - 64	64
296k	128 - 128	64
702k	128 - 128 - 128	180
3362k	64 - 128 - 256	512 - 128
TempCNN		

References

1. Bailly, A., Malinowski, S., Tavenard, R., Chapel, L., Guyet, T.: Dense bag-of-temporal-SIFT-words for time series classification. In: Douzal-Chouakria, A., Vilar, J.A., Marteau, P.-F. (eds.) AALTD 2015. LNCS (LNAI), vol. 9785, pp. 17–30. Springer, Cham (2016). https://doi.org/10.1007/978-3-319-44412-3_2
2. Bailly, S., Giordano, S., Landrieu, L., Chehata, N.: Crop-rotation structured classification using multi-source Sentinel images and LPIS for crop type mapping. In: IGARSS (2018)
3. Chung, J., Gulcehre, C., Cho, K., Bengio, Y.: Empirical evaluation of gated recurrent neural networks on sequence modeling. CoRR (2014)
4. Garnot, V.S.F., Landrieu, L., Giordano, S., Chehata, N.: Time-space tradeoff in deep learning models for crop classification on satellite multi-spectral image time series. In: IGARSS (2019)

5. Garnot, V.S.F., Landrieu, L., Giordano, S., Chehata, N.: Satellite image time series classification with pixel-set encoders and temporal self-attention. In: CVPR (2020)
6. Ji, S., Zhang, C., Xu, A., Shi, Y., Duan, Y.: 3D convolutional neural networks for crop classification with multi-temporal remote sensing images. Remote Sens. **10**, 75 (2018)
7. Kussul, N., Lemoine, G., Gallego, F.J., Skakun, S.V., Lavreniuk, M., Shelestov, A.Y.: Parcel-based crop classification in Ukraine using Landsat-8 data and Sentinel-1A data. IEEE J. Sel. Top. Appl. Earth Obs. Remote Sens. **9**, 2500–2508 (2016)
8. Pelletier, C., Valero, S., Inglada, J., Champion, N., Dedieu, G.: Assessing the robustness of random forests to map land cover with high resolution satellite image time series over large areas. Remote Sens. Environ. **187**, 156–168 (2016)
9. Pelletier, C., Webb, G.I., Petitjean, F.: Temporal convolutional neural network for the classification of satellite image time series. Remote Sens. **11**, 523 (2019)
10. Rußwurm, M., Körner, M.: Convolutional LSTMs for cloud-robust segmentation of remote sensing imagery. In: NeurIPS Workshop (2018)
11. Rußwurm, M., Körner, M.: Self-attention for raw optical satellite time series classification. arXiv preprint arXiv:1910.10536 (2019)
12. Tavenard, R., Malinowski, S., Chapel, L., Bailly, A., Sanchez, H., Bustos, B.: Efficient temporal kernels between feature sets for time series classification. In: Ceci, M., Hollmén, J., Todorovski, L., Vens, C., Džeroski, S. (eds.) ECML PKDD 2017. LNCS (LNAI), vol. 10535, pp. 528–543. Springer, Cham (2017). https://doi.org/10.1007/978-3-319-71246-8_32
13. Vaswani, A., et al.: Attention is all you need. In: NeurIPS (2017)
14. Wu, Y., He, K.: Group normalization. In: Ferrari, V., Hebert, M., Sminchisescu, C., Weiss, Y. (eds.) ECCV 2018. LNCS, vol. 11217, pp. 3–19. Springer, Cham (2018). https://doi.org/10.1007/978-3-030-01261-8_1
15. Xingjian, S., Chen, Z., Wang, H., Yeung, D.Y., Wong, W.K., Woo, W.c.: Convolutional LSTM network: a machine learning approach for precipitation nowcasting. INeurIPS (2015)
16. Ye, L., Keogh, E.: Time series shapelets: a new primitive for data mining. In: ACM SIGKDD (2009)

Creating and Characterising Electricity Load Profiles of Residential Buildings

James Fitzpatrick[1]($\boxtimes$) (iD), Paula Carroll[1] (iD), and Deepak Ajwani[2] (iD)

[1] Quinn School of Business, University College Dublin, Dublin, Ireland
james.fitzpatrick1@ucdconnect.ie, paula.carroll@ucd.ie
[2] School of Computer Science, University College Dublin, Dublin, Ireland
deepak.ajwani@ucd.ie

Abstract. Intelligent planning, control and forecasting of electricity usage has become a vitally important element of the modern conception of the energy grid. Electricity smart-meters permit the sequential measurement of electricity usage at an aggregate level within a dwelling at regular time intervals. Electricity distributors or suppliers are interested in making general decisions that apply to large groups of customers, making it necessary to determine an appropriate electricity usage behaviour-based clustering of these data to determine appropriate aggregate load profiles. We perform a clustering of time series data associated with 3670 residential smart meters from an Irish customer behaviour trial and attempt to establish the relationship between the characteristics of each cluster based upon responses provided in an accompanying survey. Our analysis provides interesting insights into general electricity usage behaviours of residential consumers and the salient characteristics that affect those behaviours. Our characterisation of the usage profiles at a fine-granularity level and the resultant insights have the potential to improve the decisions made by distribution and supply companies, policy makers and other stakeholders, allowing them, for example, to optimise pricing, electricity usage, network investment strategies and to plan policies to best affect social behavior.

Keywords: Smart-meter · Load-profiling · Time series clustering

1 Introduction

Accurately characterizing the daily load profile of electricity usage has the potential to considerably improve the decision making for electricity suppliers and distributors, customers, policy makers and various other stakeholders. For instance, it can help suppliers to optimise pricing, distributors to develop better distribution strategies, manage the peak demand and find ways to flatten the peak and it can support policy makers to align climate action plans with cleaner energy initiatives.

In particular, a careful analysis of the smart meter time series data, with a view to learn insights for characterizing the daily load profile of residential

© Springer Nature Switzerland AG 2020
V. Lemaire et al. (Eds.): AALTD 2020, LNAI 12588, pp. 182–203, 2020.
https://doi.org/10.1007/978-3-030-65742-0_13

customers has potential to assist the various stakeholders in taking a data-driven approach to their decision making. However, extracting these insights and understanding the connections between electricity usage, the dwelling and the consumer behaviour is non-trivial. The smart meter time series data at the individual dwelling level are noisy but when aggregated to groups of users evaluated over time can reveal patterns of behaviours. Such patterns, or representative load profiles, indicate when the peak demand may occur for groups of customers, and are used by electricity market operators to schedule generation to meet demand. There are opportunities to encourage users to moderate their electricity usage patterns so as to reduce aggregate peak demand, but first we need to develop an understanding of the representative load profiles.

In this paper, we consider the case when the user remains in control of their electricity usage, rather than an intelligent energy management system. We take the perspective of an electricity supplier or policy maker wishing to understand residential consumers electricity usage. We base our work on a smart meter customer behaviour trial which was carried between 2009 and 2011 [6]. Participants retained total autonomy over the scheduling of their electricity usage during the trial. For each participant, a survey was carried out before and after the installation to determine the characteristics of the building construction and the household composition, as well as their attitudes to the electricity usage and expected benefits of a smart meter. This multivariate data-set of smart meter time series and survey responses provides a unique opportunity to study the relationship between the characteristics of a dwelling and its electricity consumption pattern, when the consumption information is accessible to the users. Policy makers would be interested know which of the survey features best explain consumer electricity usage patterns. Analysis of survey responses using explainable techniques provides actionable insights that can be targeted in electricity efficiency programmes.

The smart meter usage data are stochastic and high-dimensional. In order for actors in the electricity market to incorporate these data into data-driven decision-making processes, we must consider how to reduce the dimensionality, model the data and extract useful insights. In this paper, we explore appropriate schemes for carrying this out and to answer the following research questions:

1. Can we create representative load profiles for clusters of smart meter users based on time series electricity usage?
2. Can we characterise the cluster representative load profile using information in the survey?
3. What insights do the cluster characteristics provide to support the development of climate action and electricity usage incentives?

We address these research questions by first performing a careful clustering of the normalized electricity load time-series to learn the behavioural daily patterns of residential customers. Then, we learn a classification model to map the survey features to the clusters. In the process, we focus on the importance of various survey features and learn crucial insights from this analysis. Our insights can be valuable for distribution and supply companies, policy makers and various stakeholders in the energy business. For instance, we learn that one of the most important features in predicting the daily load cluster is "how strongly they feel that they can convince other occupants of the building to reduce their energy usage". Given that the survey has many detailed characteristics of the building and the household, the consistent importance of this feature across many different classification models is surprising. This, itself, is an important finding for a country like Ireland, which has traditionally struggled to get value out of retro-fitting houses for energy efficiency improvements [16]. Our finding suggests that a marketing campaign to change the attitudes of people towards energy efficiency may be effective in modifying the daily usage pattern of residential customers.

Outline. This rest of the paper is structured as follows: Sect. 2 describes the related work, Sect. 3 details the structure of the time series and describes the survey data and the problem outline, Sect. 4 concerns the clustering of the time series and the creation of aggregate load profiles and the process of mapping survey responses to their corresponding time series cluster, Sect. 5 illustrates the experimental results and provides an exposition on these results and Sect. 6 presents our conclusions.

2   Related Work

In this section, we review the literature related to the usage of time series clustering for smart meter data. We briefly survey (i) the techniques developed for time-series clustering in general, then (ii) cover the work related to the usage of time-series clustering for smart meter electricity data with a specific focus on the Irish customer behaviour trial data and (iii) characterization of smart meter load profile of residential users based on the attributes of the residential building.

Time-Series Clustering. Clustering of time-series data has been an active area of research over the last few decades and many good techniques have been developed (c.f. [2,11,17] for surveys and [13] for some recent work). The challenge in clustering the smart meter data stems from:

1. Electricity usage time series is inherently noisy. Such noise emerges naturally from the stochasticity of human lifestyles, but also from climactic and weather conditions, and even possibly the purposeful injection of noise to ensure privacy [9]).
2. Time-series differ in length. While this challenge is typically addressed using dynamic time warping measures (as highlighted in the review [7]), these methods are sensitive to noise, making the resolution of the first challenge even more challenging.
3. We are not interested in clustering based on the total usage, but in identifying different shapes of the standardised time-series, corresponding to the different daily patterns of the consumers. The tasks of clustering based on

total usage and learning to forecast the load based on attributes of the household are relatively easier, our task of learning the daily pattern of a household is significantly more difficult.

4. For the public policy bodies and industry analysts to be able to act on the models and the resultant insights, the clustering of the time series and the mapping of the survey data to the clusters should be as interpretable as possible.

Analysis of Smart Metering Infrastructure. The installation of smart metering infrastructure in recent years has sparked interest in the desire to develop methods to draw insights from the data that is being collected. This includes not only electrical smart meters, but also water and gas smart meters [4,5,12,15]. Understanding how groups of consumers behave makes it possible to plan infrastructure projects, develop pricing strategies and identify anomalous behaviours. Naturally, clustering can be performed trivially for cases of separating commercial and industrial consumers from residential consumers, as well as by grouping by consumption magnitude. In contrast to most existing works, our focus is on the considerably more difficult task of learning the behaviour-based clusters to better understand how consumers consume. Such a clustering reveals the different daily usage patterns of residential customers and enables us to learn which features of the buildings, households and people's attitudes best discriminate between the different clusters, revealing crucial insights for policy makers.

Clustering the Time-Series Daily Usage Pattern from Household. While there is considerable body of work on clustering residential electricity customers using load time series (see e.g., [14]), there is very little work on correlating it with the features of the household and building, leave aside our goal of inferring the usage pattern from the household and building features. Lavin and Klabjan [10] constructed mean normalised daily energy profiles for each meter in their data-set of commercial and industrial buildings in the United States. They noted that the daily usage pattern could be used to determine the work schedule in the commercial buildings. Note that our focus is on the significantly more challenging task of learning the behavioural usage pattern from the household and building features. Alonso et al. [1] focused on scalable clustering of the time-series by reducing their time series representation to autocorrelation coefficients. They showed that the clusters that they obtained correlated well with the geo-demographic data related to the class and social status of individuals. In contrast, we take the study to the next level and attempt to infer the usage pattern from a range of features and identify the features that are most discriminatory. In Flath et al. [8], standard normalised daily load and weekly profiles for nine scenarios recognised by the German energy industry were computed as features from time series data. These previously known load profiles were used to perform clustering of the time-series data from a pricing perspective. However, they do not seek to explain the underlying characteristics of the buildings to which the smart meters are connected. Also, in contrast to their work, we identify the importance of each feature in identifying the usage patterns without any assumptions a priori.

Analysis of Irish Customer Behaviour Trial Data. There has also been some work on the analysis of the Irish customer behaviour trial data [6] that we use in this study. Carroll et al. [5] derived statistical features from the time series over a period of six months and attempted to solve the problem of inferring composition of a household living in a building based on the features that characterise the electricity usage behaviour of the smart meter time series. In contrast, this paper focuses on the significantly more challenging task of learning the usage behaviour from the features obtained using the associated survey.

A closely related work is that of McLoughlin et al. [12], who performed subsequence clustering of the CER [6] residential electricity smart meter time series by considering the first six months of recordings for each meter using self organising maps. However, they focused on the regression models and more crucially, ignored the features corresponding to how often the household appliances were used (only using if appliances such as washing machine were present in the household) and the attitudes of the occupants towards energy saving and metering measures. In contrast, we found that these features were the most important in discriminating between the different usage patterns of household customers. Azaza and Frederik [3] analyse the same data-set, using self-organizing maps and hierarchical methods, clustering the time series using daily mean energy usage profiles. But they only attempt to understand each cluster from an energy usage perspective, not a building composition perspective. In contrast, our study addresses the challenging task of learning the clusters of daily usage patterns from the accompanying survey data.

3   Smart Meter Characterisation and Classification Problem

In this work we are concerned with the creation of electricity load profiles for residential electricity consumers. Associating a load profile to each customer allows distributors and suppliers to anticipate expected user behaviour, plan infrastructure and targeted interaction strategies accordingly.

We first perform a clustering of the residential consumers into relatively large and roughly equal-sized clusters based on a transformation of their smart meter time series electricity usage. We then construct a mapping from the survey responses to these clusters to characterise the clusters. Finally we analyse the load profiles for these clusters and the salient survey questions to better understand the cluster behaviour and potential for targeted electricity savings interventions.

Dataset. For this study, we use a data-set [6] obtained from a customer behaviour trial that was carried out between 2009 and 2011. This trial was carried out in a range of Irish residential and commercial buildings to observe the response to the installation of smart meters. Participants retained total autonomy over the scheduling of their electricity usage during the trial. For each participant, a survey was carried out before and after the installation to determine the characteristics of the building construction, the composition of the household, as well as attitudes to the energy usage and expected benefits of a smart meter.

The trial includes 6445 participants, of which 4225 were residential participants. From these residential participants, we filtered out the ones with suspected instrumentation faults as well as those for whom incomplete survey responses could not be reasonably imputed. This resulted in a total of 3670 participants that were considered for our work. Each residential smart meter is assigned to one building, representing a single household.

The smart meter time series data was collected at a half-hour granularity, that is, the power consumed over each half hour interval for the duration of the study was recorded for each participant. This corresponds to 48 time slots per day, over the course of 535 days, a univariate uniformly-sampled sequence. Some time series, however, were incomplete, meaning that they are not all of the same length; they did not begin or terminate at the same time as those that extended over the entire duration. For each participant i, therefore, we have a real-valued vector $X_i \in \mathbb{R}^{d_i}$. The vast majority of these univariate time series have more than ten thousand elements.

For each participant i, there is a unique smart meter time series X_i as well as a unique survey response Z_i, forming a complete data-set $\mathcal{D} = \{(X_i, Z_i)\}_{i=1}^{3670}$. Each residential participant completed a survey prior to and subsequent to the eighteen month trial. For our analysis, we only retain responses from the pre-survey questionnaire and only if they concern the household composition (the number of people who live in the household), the characteristics of the building or its contents, or if they indicate the attitude of the respondent to the expected outcome of the trial. Questions that have categorical answers are one-hot encoded and questions that admit ordinal responses are normalised by the maximum possible value, or recorded value, if there is no maximum. This results in a 110-dimensional response vector $\hat{Z}_i \in R^{110}$ to be associated with each smart meter time series.

4   Methodology

Extracting clusters from the data is equivalent to finding a label y_i for each of the pairs (X_i, Z_i). In this section we outline the feature extraction methods we use to find a fixed length feature vector $\hat{X}_i$ to characterise each smart meter time series and cluster them into pairs $(\hat{X}_i, y_i)$. We then discuss how, having constructed feature vectors $\hat{Z}_i$ from the survey data, we find some model $p(y_i|\hat{Z}_i; \theta)$.

4.1   Time Series Clustering

In order to derive insights from the smart meter time series upon which decisions can be made, they must be reduced significantly in dimension. It follows that it is desirable to construct a small number of clusters for which analysis can be carried out. This amounts to using unsupervised methods to determine some mapping $f : \hat{X} \to k$, where $k \in \{\{0, 1, 2\}, \{0, 1, 2, 3\}, \{0, 1, 2, 3, 4\}\}$ and $\hat{X} \equiv \{\hat{X}_i\}_{i=1}^{3670}$.

Three major paradigms are recognised for the clustering of time series data: whole-series (raw) clustering, extracted feature clustering and model-based clustering [11]. These residential electricity usage time series, driven by stochastic variables such as local weather conditions and human activities, are subject to a significant degree of noise, making the first of these approaches undesirable for clustering. In addition, it is preferable that the clustering of the time series is easily interpretable, so that decisions made on the basis of the generated clusters are reliable, enabling the public policy bodies and analysts to act on the resultant models. It is, therefore, desirable to compute a feature representation that captures the behaviour of each time series and its peculiarities.

For each time series X_i we know the mapping $g_i : X_i \rightarrow \{0, 1, ..., 47\}^{m_i}$, where m_i is the number of days for which observations of the meter i were made. That is, we have an exact mapping between each recorded power consumption value and the time of day at which it was recorded. We also know the correspondence between each measurement and the day and year it was recorded. This allows us to construct fixed-length, representations of the load corresponding to fixed time periods. Consider, for example, that a smart meter is observed n times per day at regular intervals over a period of m days, then we can represent each measurement in a matrix $\boldsymbol{X} \in \mathbb{R}^{n \times m}$. Such a representation contains exactly the same information as the one-dimensional representation, but we can reduce it to obtain the mean energy usage per time slot according to:

$$\mathbb{R}^n \ni \hat{X}_j = \frac{1}{m} \sum_{i=1}^{m} X_{ij}.$$

We can also construct similar features, in order to take into account the differences in behaviour that can be observed during weekdays and weekends, or on a weekly/monthly basis. These representations are static and can be easily used as feature vectors for static clustering algorithms.

In this work, clustering was performed using the k-means clustering algorithm; various clustering algorithms were tested, such as agglomerative and other density-based methods, but k-means produced the most well-separated clusters, as indicated by computation of Silhouette indices. A variety of static representations of the time series data, such as those discussed in the previous paragraph, were chosen as the feature vectors upon which the clustering was performed. We proceeded with an ℓ_2 norm as a dissimilarity measure. In order to determine the number of clusters, trial clusterings were performed for three, four and five clusters, which suggested that clustering would be most appropriate with only three clusters. A relatively small number of clusters is desirable in this setting because it is convenient, for example, to have small representative customer groups when designing customer tariffs. We also found that with a higher number of clusters, the clusters themselves became less meaningful.

4.2   Survey Classification

The unsupervised clustering of the smart meter time series allows us to assign a label to each smart meter, indicating the membership of each smart meter to a electricity usage pattern clustering. These labels are then used to train a classification model in a supervised manner, to construct a mapping $h : \hat{Z} \to k$, where $\hat{Z} \equiv \{\hat{Z}_i\}_{i=1}^{3670}$, between the survey responses and the learned clusters. Constructing a mapping in this manner allows one to better understand the electricity usage patterns of a residential consumer using limited information about building characteristics. It is of interest to the electricity market to determine the most important of these features, so that targeted incentives and appropriate energy policies and climate plans can be designed.

Feature importance can be determined using wrapper methods, though these feature search methods can be computationally expensive if performed exhaustively. Instead, we perform our feature search using step backwards feature selection for the classification models. We perform the classification of the survey features using random forest classifiers and k-nearest neighbours classifiers, owing to the limited data available, their simplicity (and hence ease of interpretation), and in the case of the random forest models, so that we may also observe the feature importance values that are naturally computed during the learning process.

Classification Feature Selection. Evaluating the feature importance using wrapper methods requires some level of care. Since multiple features can correspond to a one-hot encoding of the same survey question, and since we are interested in determining the most important survey question, we must take care to ensure that the backwards greedy feature selection process selects features by greedily searching through questions rather than elements of the survey vectors. This is achieved by creating a custom *scikit-learn* estimator to implement the fitting logic and using *mlxtend* to perform the wrapper method search. For each model we perform step backwards greedy feature selection, we use five-fold cross-validation and use ROC-AUC as the scoring measure.

5 Experimental Results

All experiments were carried out on a machine with 15.5 GB of RAM, with Ubuntu 18.04 and a six core Intel® Core™ i7-9750H CPU 2.60 GHz processor. Each clustering and classification task was performed using tools from the *Scikit-Learn* Python package. Feature importance extraction was achieved using the *MLXtend* Python package. Due to limited time and a lack of code availability, it was not possible to make methodological comparison with the works of McLoughlin et al., Lavin and Klabjan, or Alonso et al. [1,10,12].

5.1 Feature Vectors

A variety of fixed-length feature vectors were constructed to test their usefulness for constructing clusters from the smart meter time series. The vector we denote by $d \in \mathbb{R}^{48}$ contained 48 elements (corresponding to the 48 half-hours in a day), each representing the mean electricity consumption in kilowatt-hours for the corresponding time-slot over the entire eighteen-month period of observation. That is, this vector represented the mean behaviour for a day over all recorded days. This vector was then normalised by dividing the value of each element by the sum of the values, so that the clustering would be agnostic of the magnitude of the electricity consumption. The vector denoted by $m \in \mathbb{R}^{108}$ contains the vectors $d_w \in \mathbb{R}^{48}$ and $d_e \in \mathbb{R}^{48}$, which are the same as d but computed only over weekdays and weekend days respectively, along with a vector $n \in \mathbb{R}^{12}$ representing the total energy usage for each month, normalised similarly. We also use the feature vector $w \in \mathbb{R}^{336}$, which contains the mean value of electricity usage for each time-slot over an entire week, representing the "typical week". Finally, we also make comparison with the statistical feature vector $s \in \mathbb{R}^{21}$ described in [5].

The survey data were normalised such that the maximum value that any element could take was unity and the minimum value was zero. The survey posed a respondent questions relating to the occupation, ages and number of residents in the house, whether they were present during the day, the age of the house, whether certain appliances were within it and how often they were used, as well as attitudes toward and expectations of the installation of the smart meter. For categorical features, such as the BER energy efficiency rating, a one-hot encoding was used. For discrete, ordinal features, their values were divided by the maximum possible value. In the case of the year of construction, this meant that the values were divided by 2009, the year that the study began, and re-scaled so that they took a minimum of zero. Such an assumption requires that new values falling outside this range much be clamped to the minimum and maximum values observed in this study. When values were unknown, they were imputed if imputation could be deemed reasonable. This resulted in a 110-dimensional vector, containing responses to the questions 200, 300, 420, 430, 43111, 4312, 4311, 4321, 4332, 433, 4352, 453, 6103, 460, 470, 4701, 471, 4801, 49002, 49004, 450, 452, 310, 401, 405, 410, and 4704. The statement of these questions and the permitted responses are given in Appendix A.

5.2   Behaviour Clusters

A number of algorithms were tested for clustering, but it was found that k-means with a ℓ_2 norm produced approximately equal-sized clusters reliably. We chose to partition the residential participants into three clusters, based on observations of cluster quality using the silhouette score. We performed the clustering on all 3670 feature vectors and obtain labels for each feature vector representation.

Figure 1 shows results for the clusters, the representative load profiles is the mean of the average daily electricity usage patterns for all members of clusters. We can see for the feature representations d and w that the produced clusters have approximately equal sizes and that the three clusters behave similarly in terms of average daily electricity usage. Differentiation between clusters is reflected in the usage curves, where one cluster exhibits strongly the expected diurnal electricity consumption pattern, where as the other shows much more consistent electricity usage throughout the day. Using the features m produces

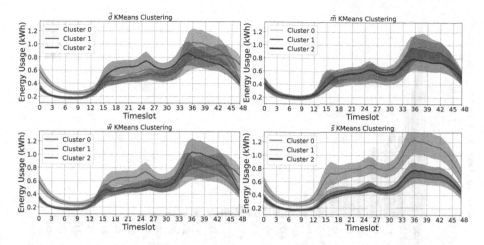

Fig. 1. Mean of the average daily electricity usage patterns for all members of clusters produced with a k means run. The shaded regions illustrate the variance of these mean values within the cluster and the thickness of the lines illustrate the relative sizes of the clusters, with the cluster having the most members represented by the thickest line.

two clusters of approximately equal size, and one smaller cluster. The behaviour of these clusters appears similar on average, but as we show later, we can establish membership of these more reliably from the information provided in the survey. The clusters produced from the features *s* demonstrate clusters that can be separated using consumption magnitude. Two of the clusters consume, in general, approximately equal magnitudes of electricity and illustrate some structural differences in their behaviour, however.

In order to assess the differentiating characteristics of each cluster, we analysed the survey responses associated with each meter. This was performed by determining a mean feature vector for each cluster and computing the variance between the mean question responses of different clusters, enabling us to identify the most discriminating questions. In Fig. 2, we illustrate this by plotting the variance across the mean question response of different clusters, which indicates the discriminating potential of different questions. We observe that for all feature vector representations, the usage rates and ownership of specific appliances turn out to be important for characterising the membership of each cluster, as indicated by questions 49001 and 49004 (see Appendix A). Interestingly, one of the most important discriminating questions is question 405, asking if the household has access to the internet or not, suggesting that users with access to internet in 2009–2011 time period had a considerably different electricity usage pattern compared to those that didn't have internet access.

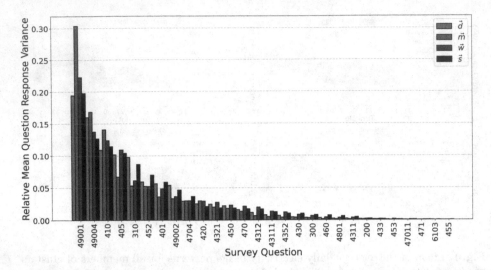

Fig. 2. Variance between the mean question response of different clusters, as a proxy to assess the discriminating potential of these survey questions in characterising the clusters – We consider the questions with high variance between their mean cluster responses as more discriminatory between clusters and the questions with similar answers across the different clusters as being less discriminatory.

5.3   Cluster Classification

Having been computed using normalised electricity usage vectors, the clusters produced are characterised by the attributes of the occupants of each building, and to a lesser extent the attributes of the building itself. This becomes further clear when we present the feature importance values based on the accompanying survey. It is possible to demonstrate which attributes these are by producing a histogram of survey responses for each cluster. In Fig. 3, we can see that cluster 0 is much more likely to respond with option 1 for the employment question, indicating that they are employed, whereas clusters 1 and 2 have a large fraction of responses with option 6, indicating that they are retired. Similarly, we can see that cluster 1 is more likely to have one or two people over the age of 15 within the building during the day time, and more likely to have fewer bedrooms.

Inspecting the characteristics of those residential buildings that have been clustered shows that the population is more likely to be distinguished by the composition of the occupants, the respondent's expectations and attitudes and the usage frequency of appliances within the residence than by the construction of the residence. In Fig. 3, we see that for clusters produced from the features m, the usage pattern corresponding to cluster one can be explained by the higher likelihood that it contains occupants who have reached pensionable age and who are less likely to have younger residents.

For the classification task, the cluster labels were used as supervised learning targets. Labels corresponding to the cluster embedding for each feature

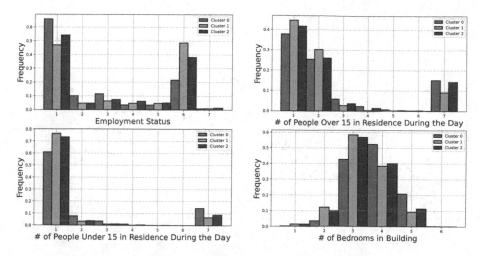

Fig. 3. Histograms illustrating the response frequencies for each cluster for select survey questions, where the clusters were constructed with the features m.

representation were tested, to determine which ones could be used to create clusters that facilitated the classification task well. The training data consisted of 75% of the participants, with the training and validation sets split evenly between the remaining 25% of the participants. The k-nearest neighbours model was tested for a variety of k values to determine the best values of $k \in \{1, 2, \cdots, 150\}$. The quality of each clustering model was determined using the testing and validation ROC-AUC and accuracy scores. In each case, the ℓ_2 norm was used as a measure of dissimilarity. Random forest models were constructed with between 100 and 1000 decision trees, using the information gain splitting technique. No maximum depth was specified and all other parameters were left as their default values according to the implementation in the *scikit-learn* package.

In Fig. 4 we evaluate the ROC-AUC score for the k nearest neighbours models on the testing sets for a variety of values of k. In each case where a valid ROC-AUC score could not be computed, a point is omitted. In general, classification accuracy is relatively low, but can be improved for larger values of k, especially when computing clusters using the features m. We note that this survey was not designed specifically for predicting the electricity usage patterns of the households and the relatively lower accuracy in our results is likely the result of the limited relevance of the survey questions to the underlying driving forces of electricity consumption profiles.

In Fig. 5 we present the testing ROC-AUC scores for a variety of forest sizes. In several cases, the scores for the clusters generated using the statistical features are best, but this is unsurprising since variables corresponding to larger buildings will allow it to make distinctions more easily. We are not interested in magnitude profiling, so we ignore the statistical feature models when evaluating feature importance.

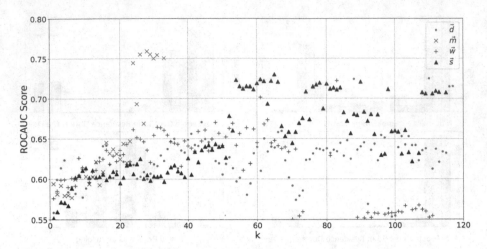

Fig. 4. ROC-AUC scores computed for k nearest neighbours classification for a range of values of k. Scores are computed for each of the feature representations.

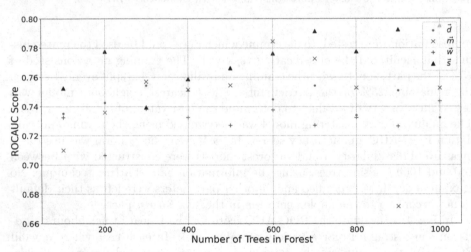

Fig. 5. ROC-AUC scores computed for the random forest classification models of various sizes. Scores are computed for each of the feature representations.

5.4   Feature Importance

Determination of the most important survey questions for correct classification of residential homes can be achieved by using a multitude of search algorithms, but performing this efficiently is difficult.

Evaluating the feature importance using wrapper methods requires some level of care. Since multiple features can correspond to a one-hot encoding of the same survey question, and since we are interested in determining the most important survey question, we must take care to ensure that the backwards greedy feature selection process selects features by greedily searching through questions rather than elements of the survey vectors. This is achieved by creating a custom *Scikit-Learn* estimator to implement the fitting logic and using *MLXtend* to perform the wrapper method search.

Feature importance values can be extracted from the random forest model implementation in *scikit-learn*. These indicate that the survey responses are dominated by few very important questions that translate to powerful features. Question 49004, is determined to be the most important, asking the respondent to indicate how often they use a variety of household appliances each day. Surprisingly, Question 4352, the next most important feature for the classification asks the participant how strongly they feel, either positively or negatively, that they can convince other occupants of the building to reduce their energy usage. The next three features included questions 49002, 49001 and 453 related to questions about how many entertainment devices of various kinds are in the home, how many household appliances of various types are in the home, and the year of construction. The most important single survey question was question 453. In Fig. 6 we can see that these five survey features remain the most important, irrespective of the features used to generate the clusters.

Performing backwards greedy feature selection for the variety of k-nearest neighbours models and random forest models outlined in the experiments above indicates that the features corresponding to these five questions are invariably the most important for classification accuracy. Although this has been computed for a limited spectrum of classification models, this suggests that these questions are, in general, the most important for classifying into the clusters constructed with relative electricity usage features.

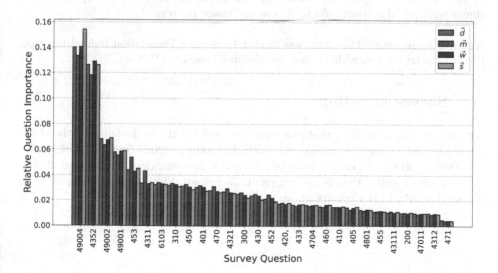

Fig. 6. Relative importance scores of survey questions, computed by the random forest classifiers. Importance values are computed for each case of the features used to determine the clusters.

6   Conclusions

In this work, we constructed a clustering of smart-meter time series for residential homes based individual average load profiles, deriving representative load profiles for the entire cluster. Using the cluster labels, we trained classification models to predict cluster membership using only occupancy, building construction and attitudinal survey responses. We identified the most relevant survey questions for performing such a classification, and those that are not, assigning relative importance values to each question obtained using random forest classifiers. We confirmed these results by performing step backwards greedy feature selection, identifying usage of appliances, age of the building and attitudes of occupants towards energy usage as some of the most important characteristics to explain energy usage patterns. Unlike previous studies, we found that one of the most important characteristics of occupants of a residential household that influences their consumption behaviour is reflected by how likely it is that they feel they can convince other occupants to reduce their electricity consumption. The fact that a feature based on attitude of the people is more crucial to determining the electricity usage patterns compared to many other features based on characterizing the household and the building has important implications for policy makers, particularly in Ireland, where the returns on retro-fitting houses (as part of the climate action plan) has been found to be very poor. Our study suggests that a marketing campaign to alter the behavioural attitudes of people might be more effective in altering the usage patterns of residential customers.

It remains to determine precisely which questions would be more effective for improving the classification accuracy. Further work could be carried out to test alternative questions that will enable us to more accurately map the characteristics of a household to its energy usage patterns.

Acknowledgement. This work was funded by Science Foundation Ireland through the SFI Centre for Research Training in Machine Learning (18/CRT/6183).

A   Survey Questions

Answers to the following questions were retained for use as features in the classification task. Note that questions 49003,1, 49003,2, 49003,3, 49003,4, 490004, 4900004, 4900005, 4900006, 4900007, 4900008, are encoded as question 49004 in the above results (that is, questions 46-55 below). Similarly, question 4551 is encoded as 455 in the results above.

1. 200 **PLEASE RECORD SEX FROM VOICE**
   - ☐ Male
   - ☐ Female

2. 300 **May I ask what age you were on your last birthday?**
   - ☐ 18 - 25
   - ☐ 26 - 35
   - ☐ 36 - 45
   - ☐ 46 - 55
   - ☐ 56 - 65
   - ☐ 65+
   - ☐ Refused

3. 310 **What is the employment status of the chief income earner in your household, is he/she**
   - ☐ An employee
   - ☐ Self-employed (with employees)
   - ☐ Self-employed (with no employees)
   - ☐ Unemployed (actively seeking work)
   - ☐ Unemployed (not actively seeking work)
   - ☐ Retired
   - ☐ Carer: Looking after relative or family

4. 401 **SOCIAL CLASS: Interviewer, Respondent said that occupation of chief income earner was....**
   - ☐ AB
   - ☐ C1
   - ☐ C2
   - ☐ DE
   - ☐ F [RECORD ALL FARMERS]
   - ☐ Refused
   - ☐ Carer: Looking after relative or family

5. 410 **What best describes the people you live with?**
   - ☐ I live alone
   - ☐ All people in my home are over 15 years of age
   - ☐ Both adults and children under 15 years of age live in my home

6. 420 **How many people over 15 years of age live in your home?**
   - ☐ 1
   - ☐ 2
   - ☐ 3
   - ☐ 4
   - ☐ 5
   - ☐ 6
   - ☐ 7 or more

7. 430 **And how many of these are typically in the house during the day (for example for 5-6 hours during the day)?**
   - ☐ 1
   - ☐ 2
   - ☐ 3
   - ☐ 4
   - ☐ 5
   - ☐ 6
   - ☐ 7 or more

8. 43111 **How many people under 15 years of age live in your home?**
   - ☐ 1
   - ☐ 2
   - ☐ 3
   - ☐ 4
   - ☐ 5
   - ☐ 6
   - ☐ 7 or more

9. **4312 And how many of these are typically in the house during the day (for exanmple for 5-6 hours during the day)?**
   ☐ 1
   ☐ 2
   ☐ 3
   ☐ 4
   ☐ 5
   ☐ 6
   ☐ 7 or more

10. **4331,3 I / we am are interested in changing the way I / we use electricity if it reduces the bill**
    ☐ 1 - strongly agree
    ☐ 2
    ☐ 3
    ☐ 4
    ☐ 5 - strongly disagree

11. **4331,4 I / we am are interested in changing the way I / we use electricity if it helps the environment**
    ☐ 1 - strongly agree
    ☐ 2
    ☐ 3
    ☐ 4
    ☐ 5 - strongly disagree

12. **4331,5 I / we can reduce my electricity bill by changing the way the people I / we live with use electricity**
    ☐ 1 - strongly agree
    ☐ 2
    ☐ 3
    ☐ 4
    ☐ 5 - strongly disagree

13. **4321,2 I / we have already done a lot to reduce the amount of electricity I / we use**
    ☐ 1 - strongly agree
    ☐ 2
    ☐ 3
    ☐ 4
    ☐ 5 - strongly disagree

14. **4321,3 I / we have already made changes to the way I / we live my life in order to reduce the amount of electricity we use.**
    ☐ 1 - strongly agree
    ☐ 2
    ☐ 3
    ☐ 4
    ☐ 5 - strongly disagree

15. **4321,4 I / we would like to do more to reduce electricity usage**
    ☐ 1 - strongly agree
    ☐ 2
    ☐ 3
    ☐ 4
    ☐ 5 - strongly disagree

16. **4321,5 I / we know what I / we need to do in order to reduce electricity usage**
    ☐ 1 - strongly agree
    ☐ 2
    ☐ 3
    ☐ 4
    ☐ 5 - strongly disagree

17. **433 Thinking about the energy reduction activities undertaken by you or your family/household, in the last year, did your efforts reduce your bills?**
    ☐ Yes
    ☐ No
    ☐ Don't know

18. 4352,2 **It is too inconvenient to reduce our usage of electricity**
    ☐ 1 - strongly agree
    ☐ 2
    ☐ 3
    ☐ 4
    ☐ 5 - strongly disagree

19. 4352,3 **I do not know enough about how much electricity different appliances use in order to reduce my usage**
    ☐ 1 - strongly agree
    ☐ 2
    ☐ 3
    ☐ 4
    ☐ 5 - strongly disagree

20. 4352,4 **I am not be able to get the people I live with to reduce their electricity usage**
    ☐ 1 - strongly agree
    ☐ 2
    ☐ 3
    ☐ 4
    ☐ 5 - strongly disagree

21. 4352,5 **I do not have enough time to reduce my electricity usage**
    ☐ 1 - strongly agree
    ☐ 2
    ☐ 3
    ☐ 4
    ☐ 5 - strongly disagree

22. 4352,6 **I do not want to be told how much electricity I can use**
    ☐ 1 - strongly agree
    ☐ 2
    ☐ 3
    ☐ 4
    ☐ 5 - strongly disagree

23. 4352,7 **Reducing my usage would not make enough of a difference to my bill**
    ☐ 1 - strongly agree
    ☐ 2
    ☐ 3
    ☐ 4
    ☐ 5 - strongly disagree

24. 450 **I would now like to ask some questions about your home. Which best describes your home?**
    ☐ Apartment
    ☐ Semi-detached house
    ☐ Detached house
    ☐ Terraced house
    ☐ Bungalow
    ☐ Refused

25. 452 **Do you own or rent your home?**
    ☐ Rent (from a private landlord)
    ☐ Rent (from a local authority)
    ☐ Own Outright (not mortgaged)
    ☐ Own with mortgage etc
    ☐ Other

26. 453 **What year was your house built**
    ☐ INT ENTER FOR EXAMPLE: 1981- CAPTURE THE FOUR DIGITS

27. 6103 **What is the approximate floor area of your home?**

28. 460 **How many bedrooms are there in your home?**
    ☐ 1
    ☐ 2
    ☐ 3
    ☐ 4
    ☐ 5 +
    ☐ Refused

29. 470 **Which of the following best describes how you heat your home?**
    - ☐ Electricity (electric central heating storage heating)
    - ☐ Electricity (plug in heaters)
    - ☐ Gas
    - ☐ Oil
    - ☐ Solid fuel
    - ☐ Renewable (e.g. solar)
    - ☐ Other

30. 47001 **Do you have a timer to control when your heating comes on and goes off?**
    - ☐ Yes
    - ☐ No

31. 4701 **Which of the following best describes how you heat water in your home?**
    - ☐ Central heating system
    - ☐ Electric (immersion)
    - ☐ Electric (instantaneous heater)
    - ☐ Gas
    - ☐ Oil
    - ☐ Solid fuel boiler
    - ☐ Renewable (e.g. solar)
    - ☐ Other

32. 47011 **Do you have a timer to control when your hot water/immersion heater comes on and goes off?**
    - ☐ Yes
    - ☐ No

33. 4801 **Do you use your immersion when your heating is not switched on?**
    - ☐ Yes
    - ☐ No

34. 4704 **Which of the following best describes how you cook in your home**
    - ☐ Electric cooker
    - ☐ Gas cooker
    - ☐ Oil fired cooker
    - ☐ Solid fuel cooker (stove aga)

35. 471 **Returning to heating your home, in your opinion, is your home kept adequately warm?**
    - ☐ Yes
    - ☐ No

36. 49001,1 **Please indicate how many of the following appliances you have in your home? Washing machine**
    - ☐ None
    - ☐ 1
    - ☐ 2
    - ☐ More than 2

37. 49001,2 **Please indicate how many of the following appliances you have in your home? Tumble dryer**
    - ☐ None
    - ☐ 1
    - ☐ 2
    - ☐ More than 2

38. 49001,3 **Please indicate how many of the following appliances you have in your home? Dishwasher**
    - ☐ None
    - ☐ 1
    - ☐ 2
    - ☐ More than 2

39. 49001,4 **Please indicate how many of the following appliances you have in your home? Electric shower (instant)**
    - ☐ None
    - ☐ 1
    - ☐ 2
    - ☐ More than 2

40. 49001,5 **Please indicate how many of the following appliances you have in your home? Electric shower (electric pumped from hot tank)**
   ☐ None
   ☐ 1
   ☐ 2
   ☐ More than 2

41. 49001,6 **Please indicate how many of the following appliances you have in your home? Electric cooker**
   ☐ None
   ☐ 1
   ☐ 2
   ☐ More than 2

42. 49001,7 **Please indicate how many of the following appliances you have in your home? Electric heater (plug-in convector heaters)**
   ☐ None
   ☐ 1
   ☐ 2
   ☐ More than 2

43. 49001,8 **Please indicate how many of the following appliances you have in your home? Stand alone freezer**
   ☐ None
   ☐ 1
   ☐ 2
   ☐ More than 2

44. 49001,9 **Please indicate how many of the following appliances you have in your home? A water pump or electric well pump or pressurised water system**
   ☐ None
   ☐ 1
   ☐ 2
   ☐ More than 2

45. 49001,10 **Please indicate how many of the following appliances you have in your home? Immersion**
   ☐ None
   ☐ 1
   ☐ 2
   ☐ More than 2

46. 49003,2 **In a typical day, how often would you or your family/household use each appliance - please think of the total use by all household/family members. Washing machine**
   ☐ Less than 1 load a day typically
   ☐ 1 load typically
   ☐ 2 to 3 loads
   ☐ More than 3 loads

47. 49003,3 **In a typical day, how often would you or your family/household use each appliance - please think of the total use by all household/family members. Tumble dryer**
   ☐ Less than 1 load a day typically
   ☐ 1 load typically
   ☐ 2 to 3 loads
   ☐ More than 3 loads

48. 49003,4 **In a typical day, how often would you or your family/household use each appliance - please think of the total use by all household/family members. Dishwasher**
   ☐ Less than 1 load a day typically
   ☐ 1 load typically
   ☐ 2 to 3 loads
   ☐ More than 3 loads

49. 490004 **In a typical day, how often would you or your family/household use each appliance - please think of the total use by all household/family members. Electric shower (instant)**
   ☐ Less than 5 mins
   ☐ 5-10 mins
   ☐ 10-20 mins
   ☐ Over 20 mins

50. 4900004 **In a typical day, how often would you or your family/household use each appliance - please think of the total use by all household/family members. Electric shower (pumped from hot tank)**
    - ☐ Less than 5 mins
    - ☐ 5-10 mins
    - ☐ 10-20 mins
    - ☐ Over 20 mins

51. 4900004 **In a typical day, how often would you or your family/household use each appliance - please think of the total use by all household/family members. Electric shower (pumped from hot tank)**
    - ☐ Less than 5 mins
    - ☐ 5-10 mins
    - ☐ 10-20 mins
    - ☐ Over 20 mins

52. 4900005 **In a typical day, how often would you or your family/household use each appliance - please think of the total use by all household/family members. Electric cooker**
    - ☐ Less than 30 mins
    - ☐ 30-60 mins
    - ☐ 1-2 hours
    - ☐ Over 2 hours

53. 4900006 **In a typical day, how often would you or your family/household use each appliance - please think of the total use by all household/family members. Electric heater (plug-in)**
    - ☐ Less than 30 mins
    - ☐ 30-60 mins
    - ☐ 1-2 hours
    - ☐ Over 2 hours

54. 4900007 **In a typical day, how often would you or your family/household use each appliance - please think of the total use by all household/family members. Water pump**
    - ☐ Less than 30 mins
    - ☐ 30-60 mins
    - ☐ 1-2 hours
    - ☐ Over 2 hours

55. 4900008 **In a typical day, how often would you or your family/household use each appliance - please think of the total use by all household/family members. Immersion water**
    - ☐ Less than 30 mins
    - ☐ 30-60 mins
    - ☐ 1-2 hours
    - ☐ Over 2 hours

56. 4551 **What rating did your house achieve?**
    - ☐ A
    - ☐ B
    - ☐ C
    - ☐ D
    - ☐ E
    - ☐ F
    - ☐ G

57. 405 **Do you have internet access in your home?**
    - ☐ Yes
    - ☐ No

References

1. Alonso, A.M., Nogales, F.J., Ruiz, C.: Hierarchical clustering for smart meter electricity loads based on quantile autocovariances. IEEE Trans. Smart Grid (2020)
2. Atluri, G., Karpatne, A., Kumar, V.: Spatio-temporal data mining: a survey of problems and methods. ACM Comput. Surv. **51**(4), 83:1–83:41 (2018)

3. Azaza, M., Wallin, F.: Smart meter data clustering using consumption indicators: responsibility factor and consumption variability. Energy Procedia **142**, 2236–2242 (2017)
4. Caroll, P., Dunne, J., Hanley, M., Murphy, T.: Exploration of electricity usage data from smart meters to investigate household composition. In: Conference of European Statisticians, Geneva, Switzerland, 25–27 September 2013 (2013)
5. Carroll, P., Murphy, T., Hanley, M., Dempsey, D., Dunne, J.: Household classification using smart meter data. J. Off. Stat. **34**(1), 1–25 (2018)
6. Commission for Energy Regulation (CER): CER Smart Metering Project - Electricity Customer Behaviour Trial, 2009–2010, 1st edn. Irish Social Science Data Archive. SN: 0012-00 (2012). www.ucd.ie/issda/CER-electricity
7. Esling, P., Agon, C.: Time-series data mining. ACM Comput. Surv. (CSUR) **45**(1), 1–34 (2012)
8. Flath, C., Nicolay, D., Conte, T., van Dinther, C., Filipova-Neumann, L.: Cluster analysis of smart metering data. Bus. Inf. Syst. Eng. **4**(1), 31–39 (2012)
9. Ghasemkhani, A., Yang, L., Zhang, J.: Learning-based demand response for privacy-preserving users. IEEE Trans. Ind. Inform. **15**(9), 4988–4998 (2019)
10. Lavin, A., Klabjan, D.: Clustering time-series energy data from smart meters. Energy Effic. **8**(4), 681–689 (2014). https://doi.org/10.1007/s12053-014-9316-0
11. Liao, T.W.: Clustering of time series data—a survey. Pattern Recognit. **38**(11), 1857–1874 (2005)
12. McLoughlin, F., Duffy, A., Conlon, M.: A clustering approach to domestic electricity load profile characterisation using smart metering data. Appl. Energy **141**, 190–199 (2015)
13. Meguelati, K., Fontez, B., Hilgert, N., Masseglia, F.: High dimensional data clustering by means of distributed Dirichlet process mixture models. In: 2019 IEEE International Conference on Big Data (Big Data), pp. 890–899. IEEE (2019)
14. Motlagh, O., Berry, A., O'Neil, L.: Clustering of residential electricity customers using load time series. Appl. Energy **237**, 11–24 (2019)
15. Mounce, S., Furnass, W., Goya, E., Hawkins, M., Boxall, J.: Clustering and classification of aggregated smart meter data to better understand how demand patterns relate to customer type. In: Proceedings of Computing and Control for the Water Industry (CCWI 2016) (2016)
16. O'Doherty, C.: Half of the homes in retrofit plan no better off despite cost, 29 April 2020. https://www.independent.ie/irish-news/half-of-the-homes-in-retrofit-plan-no-better-off-despite-cost-39166389.html (2020). Accessed 16 June 2020
17. Wang, X., Mueen, A., Ding, H., Trajcevski, G., Scheuermann, P., Keogh, E.J.: Experimental comparison of representation methods and distance measures for time series data. Data Min. Knowl. Discov. **26**(2), 275–309 (2013)

Trust Assessment on Streaming Data: A Real Time Predictive Approach

Tao Peng[✉], Sana Sellami, and Omar Boucelma

Aix-Marseille Univ, Université de Toulon, CNRS, LIS,
Laboratoire d'Informatique et Systèmes, UMR 7020, Marseille, France
{tao.peng,sana.sellami,omar.boucelma}@univ-amu.fr

Abstract. IoT data, that most often carry a temporal dimension, can be exploited from an analysis perspective or from a forecasting one. In this paper, we propose a predictive approach to address the problem of *data trustworthiness* in a data stream generated by a Smart Home application. We describe an online *Ensemble Regression* model that performs prediction in assigning a trust score to a target temporal value in real-time. Experiments conducted with data retrieved from the UCI ML repository demonstrate the performance of the model, while assessing data accuracy.

Keywords: Data trustworthiness · Smart home · Data stream

1 Introduction

Among the large spectrum of IoT applications, time-series data generated by a set of sensors and actuators are integrated to form a data stream. Smart Homes are probably the trendiest domain where data stream can be exploited in different ways such as remote control of home appliances, or even securing a house, assuming the data is reliable. Unfortunately, like any data gathered from hardware devices, sensor data stream may rise quality issues such as inaccuracy or incompleteness [21], leading to difficulties in a decision making process. Within this landscape, trusting the data is a key issue for helping stakeholders involved in such process.

Trust can be handled through the concept of *Data Trustworthiness* (DT) for which there is no unified definition in the literature: for example, [16] considers that DT assessment should be consistent with quality dimensions such as accuracy, timeliness and completeness; [18,28] highlights accuracy as a DT evaluation while [1] emphasizes on subjectivity and accuracy.

In this paper, we consider accuracy as the main quality dimension for assessing data trustworthiness in a Smart Home application, assuming that data arrives on time and the data is complete. Data accuracy, which refers to the correctness of sensor measurements [21], has been recognized as the most important dimension in several papers [1,16,18,28]. It is worth noticing that accuracy

© Springer Nature Switzerland AG 2020
V. Lemaire et al. (Eds.): AALTD 2020, LNAI 12588, pp. 204–219, 2020.
https://doi.org/10.1007/978-3-030-65742-0_14

is an objective description while DT is a subjective estimation based on some assumptions (i.e., data follows a specific probability distribution). Considering the accuracy dimension and subjectivity, we borrow DT definition from [1] that is: *"Data Trustworthiness in IoT Networks is the subjective probability that data observed by a user is consistent with the data at the source"*. Note that this definition is generic enough that leaves the door open to several implementations, depending on the context and on the probability distribution(s) one may adopt.

The remainder of the paper is organized as follows: Sect. 2 reviews some related work. In Sect. 3, we describe our approach. Section 4 illustrates the experimental results. Finally, we conclude and present some perspectives in Sect. 5.

2   Related Work

DT can be assessed by means of **data similarity** such as in [13] where authors propose a pattern-wise method: a target (sensor) value is considered as reliable if it co-occurs more frequently with the value of its neighbor sensor. However, this method is rather suitable for value states (such as 0/1 represents whether it is raining) than for continuous values (such as temperature). Won *et al.* [28] consider that if multiple sensors measure the same value of interest at different indoor locations, the difference between the measured values is proportional to the distance between sensors. DT is inversely proportional to the weighted sum of the difference between test data and neighbor sensor values: the smaller is the distance between sensors, the greater is the weight.

All the above works [13,28] make the same assumption that *similar/redundant data support each other for gaining trust*. But there aren't always redundant sensors in a smart home: for example, there may be only one humidity sensor per room.

Provenance-based methods rely on different **data lineage dimensions**. In [6], inter-dependency between five items is considered: (a) data similarity, (b) data conflict, (c) path similarity, (d) data deduction, (e) provider reputation. Authors propose an iterative process for computing a trust score: at each iteration, the trustworthiness of data and provider is adjusted according to the above five elements. This work is extended by Wang *et al.* [27] in integrating the user's feedback: data received by the user come with a 'reported trust', and the user will provide its 'adjusted trust' after accepting the data. If the difference between 'reported trust' and 'adjusted trust' is too large, the provider's reputation decreases. Lim *et al.* [22] also extended work of [6] in providing a cyclic trust computation framework suitable for data streams: (a) the more trusted data reported by the sensor, the higher is the (provider's) reputation; (b) data trust depends both on data similarity, provenance similarity and sensor reputation.

The idea behind the provenance-based approach [6,22,27] is the same: *the more a data has similar redundant data with different lineages, the more this data is trusted*. However, a Smart Home is often an Ad Hoc network [23] where there is a unique data lineage from the sensor to the gateway [21].

More recent works [1, 15] promote **regression based methods**. In [15] a static city weather data set is analysed: authors propose a method that estimates the value of a target sensor by means of the values of its surrounding sensors. If the residual between the estimation and the real value exceeds a predefined threshold, then the (target) value is considered as untrusted, the residual being the difference between the predicted value and the real value. Adams *et al.* [1] revisit the work of [15] in considering that the residual follows a Gaussian distribution. A Cumulative Distribution Function takes the residual as input and outputs a trust score: if this score exceeds a threshold, the received data is trusted. Work in [1] shows that Linear Regression outperforms Random Forest Regression, Gradient Boosted Machine and Multi-Layer Perceptrons.

These works [1, 15] share the idea that *a small residual (i.e., the model made a good prediction) leads to a high trust score.*

We found the approach described in [1] appealing although it does not take into account data stream characteristics (timeliness, non-stationarity, etc.). Especially, due to seasonal changes, or changes in user habits, the underlying distribution parameters (e.g., means, variance, correlation) of smart home data usually changes over time, which is called the non-stationarity feature of the data stream [26]. Non-stationarity of the data stream leads to a significant degradation of the performance of the prediction/classification model, which is known as concept drift. Although the work of [1] does not take into account the non-stationarity of data stream and the concept drift, we believe it is a good start assuming we could transpose it to target (IoT) data streams.

In the next section, we describe DTOM, a Data Trustworthiness Online Method to evaluate a trust score of (a batch of) data in a real-time data stream. DTOM is based on the work [1] but differs by the following points: (1) DTOM is based on an Online Ensemble Regression model which is suitable for the analysis of online streams; (2) DTOM has a heuristic update strategy: Updated using the data from the top 50% of trust rankings per batch, and (3) DTOM has been evaluated with various real inaccurate data ratios while [1] use a (simulated) inaccurate fixed data ratio.

3   Data Trustworthiness Online Model: DTOM

In this section, we first provide a problem statement as well as algorithmic details, then we describe the Online Ensemble Regression methods we adopted.

3.1   Problem Statement

Given f sensors, each sensor generates a value within a fixed period of time. A value $d_{f',t'}$ is emitted by a sensor f', at time t'. If $d_{f',t'}$ has quality (accuracy dimension) issue, its accuracy level $da_{f',t'}$ is 0, otherwise, it is 1. Our model will give an estimated DT $dts_{f',t'} \in [0,1]$ (denoted as a Trust Score) by Eq. 1 from [1]. Estimation of $dts_{f',t'}$ is the solution to problem minimizing $|da_{f',t'} - dts_{f',t'}|$. So, the problem of assessing DT can be considered as a **Prediction** problem.

$$dts_{f',t'} = F(dr_{f',t'}, \mu, \sigma) =$$

$$
\begin{cases}
\frac{2}{\sigma\sqrt{2\pi}} \int_{-\infty}^{dr_{f',t'}} EXP\left(-\frac{(x-\mu)^2}{2\sigma^2}\right) dx, \text{if } dr_{f',t'} < \mu \\
\frac{2}{\sigma\sqrt{2\pi}} \int_{dr_{f',t'}}^{+\infty} EXP\left(-\frac{(x-\mu)^2}{2\sigma^2}\right) dx, \text{if } dr_{f',t'} \geq \mu
\end{cases}
\tag{1}
$$

3.2 Design and Implementation

DTOM approach consists of three processes: initialization (offline phase), assessment (online phase) and update (online phase).

Initialization: Given a sensor f', its historic data is noted as $Y_{f'}$, and its reference data (gathered from other sensors) is noted as $X_{f'}$. $Y_{f'}$ and $X_{f'}$ are used to initialize the ensemble Regression model (line 1, Algorithm 1). We calculate the estimation $\hat{d}_{f',t'}$ of each historical data $d_{f',t'}$. Then, we calculate the residual $dr_{f',t'}$ between $\hat{d}_{f',t'}$ and $d_{f',t'}$ (line 2). The average (resp. standard deviation) of the residual is denoted μ (resp. σ) (line 3, 4).

Algorithm 1. DTOM Initialization

Input: historic data of a sensor f', $Y_{f'}$; reference data of $Y_{f'}$, $X_{f'}$;
Output: the ensemble regression model Reg; the average of residuals, μ; the standard deviation of residuals, σ.
1: an ensemble regression model reg is initialized with $Y_{f'}$ and $X_{f'}$.
2: $setResiduals \leftarrow$ the training error (residual) of Reg with $Y_{f'}$ and $X_{f'}$
3: $\mu \leftarrow$ average of $setResidual$
4: $\sigma \leftarrow$ Standard deviation of $setResidual$
5: **Return** μ, σ, Reg;

Assessment: One data $d_{f',t'}$ arrives at a processing device (e.g., gateway) as defined in Algorithm 2 (lines 2–5). The ensemble regression generates an estimation $\hat{d}_{f',t'}$ (line 3) and gets the corresponding value of residual (line 4). The corresponding trust score $dts_{f',t'}$ is calculated by Eq. 1 from [1] (line 5).

Update: the new data $d_{f',t'}$ is also buffered, with its reference data $ref_{f',t'}$ and its trust score $dts_{f',t'}$ (line 6 in Algorithm 2). When the buffer is full (lines 7–18), the data from the top 50% of trust rankings in the buffer is used to update the regression (line 8), and the buffer is cleared (line 9).

3.3 Online Ensemble Regression

Online Ensemble Regression methods are suitable to our context especially for handling concept drifts [8,10,20]. Online Ensemble Regression is a set of individual regression models whose predictions are combined to predict new incoming

Algorithm 2. DTOM Assessment and Update

Input: New data from sensor f' at time t', $d_{f',t'}$; the reference data of $d_{f',t'}$, $ref_{f',t'}$; ensemble regression model, Reg, $Reg.predict$ is the prediction function of Reg; the residual follows a Gaussian Distribution, $N(\mu, \sigma^2)$; A buffer is used to store the data in each batch, and the upper limit of its capacity is also equal to the batch size, noted as $bufferSize$.

Output: trust score of $d_{f',t'}$, $dts_{f',t'} \in [0,1]$.

1:    $myBuffer \leftarrow \phi$ // The buffer cache is empty
2:    **if** new data $d_{f',t'}$ is generated **then**
3:       $\hat{d}_{f',t'} \leftarrow Reg.predict(ref_{f',t'})$ // generates an estimation of $d_{f',t'}$
4:       $dr_{f',t'} \leftarrow d_{f',t'} - \hat{d}_{f',t'}$ // get the residual
5:       $dts_{f',t'} = F(dr_{f',t'}, \mu, \sigma)$ // as Equation 1
6:       $myBuffer \leftarrow myBuffer \cup (d_{f',t'}, ref_{f',t'}, dts_{f',t'})$ //new data, its reference data and its trust score are buffered
7:       **if** $|myBuffer| \geq bufferSize$ **then** // when the buffer is full
8:         $Reg \leftarrow Reg$ update with the data from the top 50% of trust rankings in $myBuffer$
9:         $myBuffer \leftarrow \phi$ //the buffer is cleared
10:      **end if**
11: **end if**
12: **Return** $dts_{f',t'}$;

---

instances in real time. There are several online regression models in the literature [2–4, 7, 12, 14, 17, 19, 24, 25, 29].

Online ensemble regression methods may adopt the following strategies to accommodate concept drift (the strategies chosen for each model are shown in Table 1):

- M1) Modification of basic models' weights: The better the performance of the basic model in the latest data, its voting weight increases, otherwise the weight decreases.
- M2) Modification of basic models' parameters: If the basic model is updatable, its parameters are adjusted with new data.
- M3) Modification of basic models' parameters: If the loss of the entire ensemble regressions exceeds a threshold, new basic models are added to improve performance.
- M4) Modification of basic models' parameters: Poorly performing or too old basic models are removed to reduce the computational burden.
- A1) Selecting instances: Incorrectly predicted data is used to update the model because it may represent the trend of data changes.
- A2) Weighting instances: Incorrectly predicted data gets more weight that affect the model update.

As illustrated in Table 1, Online Ensemble Regression methods can be updated in 1) using a single piece of data (denoted "simple") or 2) multiple pieces (denoted "batch"). In terms of "explanatory", Online Ensemble Regression methods can be divided into two categories [25]: 1) implicit, online regression model does not use detection techniques of a concept drift, but is continuously updated with new4 data; 2) explicit, the update mechanism is triggered only when the concept drift is confirmed by the concept drift detection module. Some Online Ensemble Regression methods use a sliding window, while others don't (see Table 1). Most methods limit the number of base models except for [29].

In order to choose an Ensemble method, we adopted the following criteria (models that meet the criteria are marked with * in Table 1):

- **No re-accessible historic data** is one of the main differences between data streams and static data [20]. Data is accessed only once and then discarded to limit memory and storage space usages [9].
- **Batch-by-Batch update** has better stability than instance-by-instance [3,8] and is less sensitive to inaccurate data [3].
- **Implicit method** is more suitable than the explicit one (such as concept detection) in noisy data streams [20], because the latter may cause too many false alarms [8,20].
- **Limited number of basic models** reduce the storage burden [20].

As illustrated in Table 1, **AddExp** [19] and **B-NNRW** [3] are the methods that meet our criteria. **AddExp** uses a loss bound to obtain the error model, and adjusts the expert's weights according to their actual losses (M1). Each expert updates upon new arrival data (M2). If the overall performance (loss bound) is below (above) a predefined threshold, a new expert is added (M3). The pruning strategy is weakest-first or oldest-first (M4).

Note that the original version of AddExp was designed to update instance-by-instance, but AddExp can be easily extended to update Batch-by-Batch [25]. The original AddExp does not reveal which instances should be taken for training a new basic model [5]. However, for "Batch-by-Batch update", it can train/initialize a new basic model by Boosting/Bagging the instances in the current batch (A1, A2). One limitation of AddExp is that its predictions are in $[0, 1]$ interval. Another limitation is that it depends on a number of hyper-parameters, as follows: 1) factor of decreasing weight β: the weight of basic model is updated as $\omega_{t+1,i} = \omega_{t,i}\beta^{|\xi_{t,i}-y_t|}$; 2) loss required to add new expert τ: if $|\hat{y}_t - y_t| > \tau$, a new expert is added; 3) factor of new expert weight γ: the weight of the new basic model is equal to $\gamma \sum_{i=1}^{N_t} \omega_{t,i} |\xi_{t,i} - y_t|$. Where, $\omega_{t,i}$ is the weight of basic model i in time t; y_t is the dependent variable; $\xi_{t,i}$ is the estimation of y_t by basic model i; N_t is number of overall experts; $\hat{y}_t$ is the estimation of AddExp (over all basic models).

B-NNRW, a Boosting/Bagging ensemble method is based on NNRW (A1, A2), a Neural Network with Random Weights where the weights between the input layer and the hidden layer are fixed. NNRW does not update and adopts a linear assumption. Therefore, B-NNRW also adopts the linear assumption and adjusts its weights according to their loss in the last batch of data (M1). Pruning (M3) and adding (M4) basic models are also used to maintain the performance of the whole system. B-NNRW also relies on some hyper parameters such as 1) the pruning rate q: only Q models with the lowest error are eligible to participate in

Table 1. Online regression methods with their characteristics

Method	Description	New data size	Explanatory	Sliding window	Adaption	# basic model
*AddExp [19]	Additive expert ensembles regression	*both	*implicit	*no	M1, M2, M3, M4	*limited
ILLSA [17]	Incremental Local Learning Soft Sensing Algorithm	*batch	*implicit	yes	M1, M3, M4	*limited
OWE [25]	On-line Weighted ensembles regression	simple	*implicit	yes	M1, M3, M4, A1	*limited
R-FIMT-DD [14]	Ensemble of Incremental Hoeffding-based trees	simple	explicit	*no	M2, A1	*limited
AMRules [2]	Ensemble of randomized adaptive model rules	simple	explicit	*no	M3, M4	*limited
DOER [24]	Dynamic and Online Ensemble Regression	simple	*implicit	yes	M1, M2, M3, M4	*limited
VHPRE [4]	Vertical and Horizontal Partitioning for Data Stream Regression Ensemble	simple	explicit	*no	M1, M3, M4	*limited
ARF-Reg [12]	Adaptive Random Forest (ARF) for regression	simple	explicit	*no	M2, M3, M4	*limited
Online-DNNE [7]	Neural network ensembles with random weights based	simple	*implicit	*no	M2	*limited
*B-NNRW [3]	Neural network ensembles with random weights + Bagging/ Boosting	*batch	*implicit	*no	M1, M3, M4, A1, A2	*limited
Learn.++ R2C [29]	Learn++.NSE [8] + R2C	*batch	*implicit	*no	M1	not limited

the final prediction, $Q = q * M$, M being the maximum number of basic models; 2) the replacement rate r: the number of new added models is $r * M$.

4 Experimentation

4.1 Experimental Dataset

We conducted our experiments with the Appliances Energy Prediction dataset retrieved from the UCI Machine Learning Repository data portal[1] consisting of the following attributes: energy assumption, humidity and temperature. For illustration purposes, we focus on the *RH2 sensor* which is a humidity sensor in a living room area.

Dataset Volume and Velocity: One humidity sensor and one temperature sensor are installed in each room and outside the building (18 sensors in total). Data were averaged for 10 min period and gathered during 4.5 months (from 11/01/2016 to 05/27/2016) resulting in a total 12 MB CSV file with 19735 instances.

Simulated Untrusted Data (SUTD): Variance Fault (Gaussian noise) is one of several types of faults that can be injected into a data stream (randomly selected original data) to represent untrustworthy data [11]. [1] shows that the detection of Variance Fault is more difficult than others, such as Stuck Fault (replaces the true data value with a constant value), Offset (adds an a constant value to the true data value). The percentage of noisy data injected into original data (OD) varies from 5% to 65% (by step of 5%) for each experiment. Based on [1], we define SUTD as follows: $SUTD = OD + N(0, \delta')$, where N is a Gaussian distribution and δ' is the Standard Deviation of RH2 sensor data. Due to space limitations, Fig. 1 (b–e) shows RH2 data for the first 24 h, respectively without and with 5%, 35% and 65% of noise.

RH2 Sensor Data: Figure 1(a) displays RH2 data with some concept drifts detected by Page-Hinckley Test[2]. We observe that from January to March, data changes are relatively flat compared to April, May. Correspondingly, the concept drift from January to March is less visible than for April, May. This smart home sensor data with non-stationary nature (concept drift) will be used to test whether DTOM can handle the concept drift to correctly assess DT in the non-stationary data stream.

Reference Data: For RH2 sensor, the reference dataset is the data sent from other 17 humidity/temperature sensors (not including energy assumption data), and these sensors always generate correct data. Due to space limitations, Fig. 2 (a–d) shows an excerpt of RH6, T6, RH5, T5 sensor data with their statistical description.

---

[1] UCI https://archive.ics.uci.edu/ml/datasets/Appliances+energy+prediction.

[2] Details about Page-Hinkley method for concept drift detection are available at https://scikit-multiflow.github.io/scikit-multiflow/.

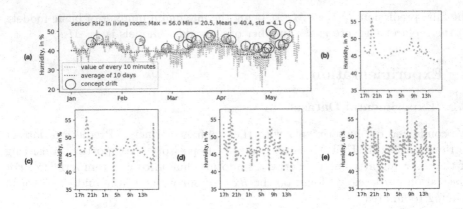

Fig. 1. (a) RH2 data in living room, from January to May, 2016. Some concept drifts are detected by Page-Hinckley Test, illustrated in circles. The first 24 h data in RH2 (b) without SUTD; with (c) 5% of SUTD; (d) 35% of SUTD; (e) 65% of SUTD.

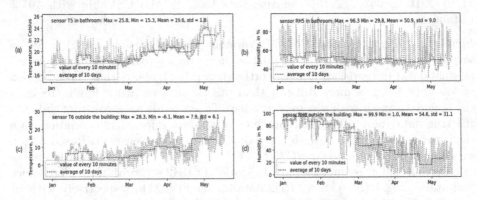

Fig. 2. (a) Temperature sensor RH5 and (b) Humidity sensor T5 in bathroom; (c) Temperature sensor RH6 and (d) Humidity sensor T6 outside the building.

Root Mean Square Error (RMSE). Equation 2 is a known measure that we use to assess OD trust. The lower the RMSE value of ODs/SUTDs, the more accurately their trust scores are estimated.

$$RMSE = \begin{cases} \sqrt{\frac{1}{|ODs|} \sum_{d_{f',t'} \in ODs} (dts_{f',t'} - 1)^2}, & \text{for ODs} \\ \sqrt{\frac{1}{|STUDs|} \sum_{d_{f',t'} \in STUDs} (dts_{f',t'} - 0)^2}, & \text{for STUDs} \end{cases} \tag{2}$$

Balanced-Accuracy (BACC). To further evaluate DTOM, data are classified either as trustworthy or untrustworthy according to a threshold *tth*. Let us set up: an OD is seen as a **true positive (TP)** if it is correctly classified as 'trustworthy' and a **false negative (FN)** otherwise, and that a SUTD is seen as a **true negative (TN)** if it is correctly identified as 'untrustworthy' and a

false positive (FP) otherwise. In this case, $BACC$ (Eq. 3) indicates whether the overall data is well classified and takes into account the unbalanced nature of the dataset.

$$BACC = (Sensibility + Specificity)/2$$

$$\text{where } Sensibility = \frac{\#TP}{\#TP + \#FN} \text{ and } Specificity = \frac{\#TN}{\#TN + \#FP} \quad (3)$$

Trust Score. We can also directly observe the trust score of ODs/SUTDs to determine whether they are correctly scored when the concept drift occurs. The expected trust score for any OD is 1. Therefore, in the case of concept drift, the higher (more accurate) of ODs' trust score, DTOM adapts better to the concept drift. Similarly, the expected trust score for any SUTD is 0. In the case of concept drift, the lower (more accurate) SUTDs' trust score, DTOM adapts better to the concept drift.

4.2   Evaluation

In order to evaluate DTOM, we implemented AddExp and B-NNRW. We also compared DTOM with linear regression (a static model), to show how DTOM behaves in presence of concept drift. For any regressor, the first 5% data are used for initialization. Trust threshold tth is determined by maximizing BACC. For any Online Ensemble Regression: the maximum number of basic models is 25; instances are weighted by Boosting; buffer size is 100. The super-parameters for each regressor are as follows:

- *AddExp*: factor of decreasing weight $\beta = 0.5$, factor of new expert weight $\gamma = 0.1$, loss required to add new expert $\tau = 0.05$ (see definitions in Sect. 3.3). These super-parameter settings are the optimal values after tuning, i.e., the same settings suggested in [19]. Basic regression models are SGD-Regressor and Passive-Aggressive-Regressor[3]. Pruning strategy is the worst first [3].
- *B-NNRW*: Number of hidden nodes of NNRW is 16; the pruning rate $p = 0.9$ (optimal value between 1.0 and 0.7); the replacement rate $r = 0.1$ (optimal value between 0.0 and 0.3) (see definitions in Sect. 3.3).
- *Linear Regression*: the first 5% data are taken for initialization; the trust threshold tth is determined by maximizing BACC, but without update.

4.3   Results

In this subsection, we will discuss the numbers obtained for RMSE (trust score's accuracy) and BACC (OD/SUTDs' classification) for all the above methods.

---

[3] Available in Sklearn: https://scikit-learn.org/stable/.

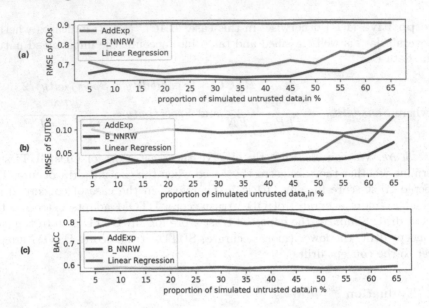

Fig. 3. (AddExp vs. B-NNRW vs. Linear Regression) performance with different SUTD ratios (5%–65%) in RH2 data: (a) RMSE of ODs' trust score; (b) RMSE of STUDs' trust score; (c) BACC of overall data.

RMSE of ODs: As depicted in Fig. 3(a), the linear regression curve is close to 0.9 for all different SUTD ratios. This means that with linear regression, OD is always wrongly evaluated with a relatively low trust score. The reason is that the residuals between ODs and their prediction are unexpectedly too large. A further explanation is that linear regression without updates cannot maintain predictive power in non-stationary data streams, due to concept drift.

Note that for all SUTD ratios, B-NNRW curve is always lower, and therefore better than AddExp. One possible explanation is that, for non-stationary data stream, the prediction ability of B-NNRW is better than AddExp one.

For a SUTD ratio in the 5%–45% range, both B-NNRW and AddExp curves are stable. In other words, B-NNRW and AddExp maintain their performance as data quality declines. The reason is that 1) DTOM has successfully filtered out most of low-quality data that is not used to update the Ensemble Regression models, 2) B-NNRW and AddExp both have a certain tolerance for inaccurate data.

However, when the SUTD ratio exceeds 50%, B-NNRW and AddExp curves increase. This is because they both are updated by using the data from the top 50% of trust rankings in each batch. If the SUTD ratio is close to or higher than 50%, SUTDs inevitably interfere with its update process.

RMSE of SUTDs: Figure 3(b) shows that, for all SUTD ratios, SUTDs' RMSE of Linear Regression is close to 0.1.

Figure 3(b) also shows that AddExp and B-NNRW performances are stable when the SUTD ratio is within a 5% to 50% range, and B-NNRW ratio (which is close to 0.02) is slightly lower (better) than AddExp (close to 0.03). Both AddExp and B-NNRW behave better than linear regression.

When the proportion of SUTD is greater than 50%, AddExp curve increases significantly. This means that AddExp loses performance: it is even worse than Linear Regression. However, B-NNRW curve increases more slowly than AddExp. One possible explanation is: 1) AddExp loses its predictive ability due to updating with some SUTDs; 2) B-NNRW has a higher tolerance than AdddExp for SUTD, and its prediction ability is less negatively impacted.

BACC: Figure 3(c) shows that, in all cases, Linear Regression BACC is stably close to 0.55, which is lower than others. This means that nearly half of the data is correctly classified.

For all SUTD ratio range values, we have shown that B-NNRW performs better than AddExp, in comparison with ODs' RMSE and of STUDs' RMSE. Therefore, BACC of B-NNRW is always higher (better) than AddExp. This means that, a higher percentage of data is correctly classified with B-NNRW than with AddExp.

From the 50%–65% SUTDs ratio, we showed that B-NNRW and AddExp have lost performance in both ODs' RMSE and STUDs' RMSE, due to unavoidable update with SUTDs. Therefore, as the SUTD ratio increases from 50%, both B-NNRW and AddExp BACC values decrease.

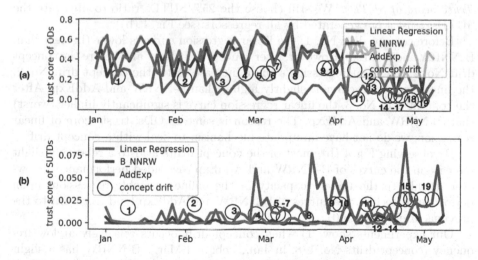

Fig. 4. (AddExp vs. B-NNRW vs. Linear Regression) performance with RH2 data (25% SUTD ratios), in case of concept drift: (a) trust score of ODs; (b) trust score of SUTDs. Up to 19 concept drifts were detected by Page-Hinckley Test (illustrated by numbers and circles).

Trust Score of ODs. We have shown that 1) when SUTDs ratio does not exceed 50%, the performance of B-NNRW and AddExp are stable; 2) with any SUTDs ratio (5%–65%), the performance of Linear Regression is stable.

For illustration purposes, we arbitrarily choose a ratio of 25% SUTDs from 0% to 50%, and illustrate the ODs' trust score generated by all regressors, as illustrated in Fig. 4 (a). Before concept drift No.1, the linear regression curve is even higher (better) than B-NNRW and AddExp ones. However, after concept drift No.1, the linear regression curve is always close to 0.1, which is far from the expected value of 1 for ODs. The reason is that the concept drift affects the performance of Linear Regression because this method does not handle concept drifts.

By observing Fig. 4 (a), both of B-NNRW or AddExp follow a downward trend due to the concept drift. The decline of the curve means that performance is reduced. However, after the performance degradation, the curves of both B-NNRW and AddExp tend to return to the previous level. This ability comes from the update process of Online Ensemble Regression, which enables DTOM to deal with the concept drift.

In comparing B-NNRW with AddExp curves, we note that 1) when there is no concept drift, the curves of both are closed; 2) when the concept drift occurs, the curve of B-NNRW declines slightly than the AddExp one: this is illustrated in Fig. 4 (a) concept drifts 1–3, 5–8, 10–19). This means that B-NNRW can adapt to changes in the data stream more quickly than AddExp, and outperforms AddExp (the same result is shown in Fig. 3 (a) with 25% STUDs ratio).

Trust Score of SUTDs. We still choose the 25% SUTDs ratio to illustrate the SUTDs' trust score generated by all regressors (see Fig. 4 (b)).

Before concept drift No. 1, the linear regression curve is lower (better) than B-NNRW and AddExp. However, after concept drift No. 1 and before the concept drift No. 2, the linear regression's curve increase. After the concept drift No. 2, the linear regression curve is slightly higher than B-NNRW and AddExp. After the concept drift No. 3, the linear regression curve is significantly higher (worst) than B-NNRW and AddExp. The reason is same as ODs' trust score of linear regression: we do not have an update mechanism to deal with a concept drift.

By observing Fig. 4 (b), most of the concept drifts lead to a relatively slight increase in the curves of B-NNRW and AddExp (degraded performance). However, thanks to the update capacity of the online ensemble regression, upon performance decline, the curves of B-NNRW and AddExp tend to return to the previous level (close to 0).

Our experiments show: 1) when concept drift occurs relatively at low frequency (concept drifts No. 1–8, in Jan., Feb. and Mar.), B-NNRW has a slight advantage over AddExp (B-NNRW 0.001 vs. AddExp 0.003 in mean); 2) When the frequency of concept drift occurs at a higher frequency (concept drifts No. 10–19, in Apr. and May), the curves of both increase (worst), but B-NNRW keeps its advantage over AddExp (B-NNRW 0.003 vs. AddExp 0.004 in mean). Hence, the overall performance of B-NNRW is better than AddExp (the same result has been shown in Fig. 3 (b) with 25% STUDs ratio).

5   Conclusion

In this article we described DTOM, an online model-based method for assessing data trustworthiness in smart home (IoT) data streams. DTOM extends the work of [1] in using Online Ensemble Regression, and in adopting a heuristic update strategy: batch-by-batch, with the data from the top 50% of trust rankings in each batch. DTOM has been implemented with B-NNRW and AddExp and experimental results have been conducted with a real dataset.

The first outcome of the experimentation is that B-NNRW ensures Data Trustworthiness for a vast majority of data in a non-stationary data stream, while outperforming other regressors. The second outcome relates to DTOM performance degradation when the SUTD ratio exceeds 50%, because SUTDs will inevitably interfere with the regressor update process.

The work described in this paper is a first step towards developing efficient real-time predictive methods for a data stream, i.e., the proposal of learning methods that (1) can handle the drifts, and (2) cover a comprehensive set of practical applications. However, the proposed methods have some limitations. Indeed, our work is based on the assumption that the initialization phase has a high-trust dataset. If a low-trust dataset is used during the initialization phase, it is possible that 1) the distribution parameters of residuals may be incorrectly estimated; 2) the parameters of the online Ensemble Regression model also may be erroneous. Clearly there is room for improving these methods. One possible research direction is that our proposed method requires only a small amount of high-trust data for initialization. This amount may be provided by domain experts at limited cost.

References

1. Adams, S., Beling, P.A., Greenspan, S., Velez-Rojas, M., Mankovski, S.: Model-based trust assessment for Internet of Things networks. In: 2018 17th IEEE International Conference on Trust, Security and Privacy in Computing and Communications/12th IEEE International Conference on Big Data Science and Engineering (TrustCom/BigDataSE), pp. 1838–1843. IEEE (2018)
2. Almeida, E., Ferreira, C., Gama, J.: Adaptive model rules from data streams. In: Blockeel, H., Kersting, K., Nijssen, S., Železný, F. (eds.) ECML PKDD 2013. LNCS (LNAI), vol. 8188, pp. 480–492. Springer, Heidelberg (2013). https://doi.org/10.1007/978-3-642-40988-2_31
3. de Almeida, R., Goh, Y.M., Monfared, R.P., Steiner, M.T.A., West, A.: An ensemble based on neural networks with random weights for online data stream regression. Soft Comput. 24(13), 9835–9855 (2020)
4. Barddal, J.P.: Vertical and horizontal partitioning in data stream regression ensembles. In: 2019 International Joint Conference on Neural Networks (IJCNN), pp. 1–8. IEEE (2019)
5. Barddal, J.P., Gomes, H.M., Enembreck, F.: Advances on concept drift detection in regression tasks using social networks theory. Int. J. Nat. Comput. Res. (IJNCR) 5(1), 26–41 (2015)

6. Dai, C., Lin, D., Bertino, E., Kantarcioglu, M.: An approach to evaluate data trustworthiness based on data provenance. In: Jonker, W., Petković, M. (eds.) SDM 2008. LNCS, vol. 5159, pp. 82–98. Springer, Heidelberg (2008). https://doi.org/10.1007/978-3-540-85259-9_6

7. Ding, J., Wang, H., Li, C., Chai, T., Wang, J.: An online learning neural network ensembles with random weights for regression of sequential data stream. Soft Comput. **21**(20), 5919–5937 (2016). https://doi.org/10.1007/s00500-016-2269-9

8. Elwell, R., Polikar, R.: Incremental learning of variable rate concept drift. In: Benediktsson, J.A., Kittler, J., Roli, F. (eds.) MCS 2009. LNCS, vol. 5519, pp. 142–151. Springer, Heidelberg (2009). https://doi.org/10.1007/978-3-642-02326-2_15

9. Ramírez-Gallego, S., Krawczyk, B., García, S., Wozniak, M., Herrera, F.: A survey on data preprocessing for data stream mining: current status and future directions. Neurocomputing **239**, 39–57 (2017)

10. Gama, J.: Knowledge Discovery from Data Streams. CRC Press, Boca Raton (2010)

11. Ganeriwal, S., Balzano, L.K., Srivastava, M.B.: Reputation-based framework for high integrity sensor networks. ACM Trans. Sens. Netw. (TOSN) **4**(3), 1–37 (2008)

12. Gomes, H.M., Barddal, J.P., Ferreira, L.E.B., Bifet, A.: Adaptive random forests for data stream regression. In: ESANN (2018)

13. Gwadera, R., Riahi, M., Aberer, K.: Pattern-wise trust assessment of sensor data. In: 2014 IEEE 15th International Conference on Mobile Data Management, vol. 1, pp. 127–136. IEEE (2014)

14. Ikonomovska, E., Gama, J., Džeroski, S.: Online tree-based ensembles and option trees for regression on evolving data streams. Neurocomputing **150**, 458–470 (2015)

15. Javed, N., Wolf, T.: Automated sensor verification using outlier detection in the Internet of Things. In: 2012 32nd International Conference on Distributed Computing Systems Workshops, pp. 291–296. IEEE (2012)

16. Jayasinghe, U., Otebolaku, A., Um, T.W., Lee, G.M.: Data centric trust evaluation and prediction framework for IoT. In: 2017 ITU Kaleidoscope: Challenges for a Data-Driven Society (ITU K), pp. 1–7. IEEE (2017)

17. Kadlec, P., Gabrys, B.: Local learning-based adaptive soft sensor for catalyst activation prediction. AIChE J. **57**(5), 1288–1301 (2011)

18. Karthik, N., Ananthanarayana, V.: Data trust model for event detection in wireless sensor networks using data correlation techniques. In: 2017 Fourth International Conference on Signal Processing, Communication and Networking (ICSCN), pp. 1–5. IEEE (2017)

19. Kolter, J.Z., Maloof, M.A.: Using additive expert ensembles to cope with concept drift. In: Proceedings of the 22nd International Conference on Machine learning, pp. 449–456 (2005)

20. Krawczyk, B., Minku, L.L., Gama, J., Stefanowski, J., Woźniak, M.: Ensemble learning for data stream analysis: a survey. Inf. Fusion **37**, 132–156 (2017)

21. Leonardi, A., Ziekow, H., Strohbach, M., Kikiras, P.: Dealing with data quality in smart home environments—lessons learned from a smart grid pilot. J. Sens. Actuator Netw. **5**(1), 5 (2016)

22. Lim, H.S., Moon, Y.S., Bertino, E.: Provenance-based trustworthiness assessment in sensor networks. In: Proceedings of the Seventh International Workshop on Data Management for Sensor Networks, pp. 2–7 (2010)

23. Lin, H., Bergmann, N.W.: IoT privacy and security challenges for smart home environments. Information **7**(3), 44 (2016)

24. Soares, S.G., Araújo, R.: A dynamic and on-line ensemble regression for changing environments. Expert. Syst. Appl. **42**(6), 2935–2948 (2015)
25. Soares, S.G., Araújo, R.: An on-line weighted ensemble of regressor models to handle concept drifts. Eng. Appl. Artif. Intell. **37**, 392–406 (2015)
26. Tran, L., Fan, L., Shahabi, C.: Outlier detection in non-stationary data streams. In: Proceedings of the 31st International Conference on Scientific and Statistical Database Management, pp. 25–36. ACM (2019)
27. Wang, X., Govindan, K., Mohapatra, P.: Provenance-based information trustworthiness evaluation in multi-hop networks. In: 2010 IEEE Global Telecommunications Conference GLOBECOM 2010, pp. 1–5. IEEE (2010)
28. Won, J., Bertino, E.: Distance-based trustworthiness assessment for sensors in wireless sensor networks. NSS 2015. LNCS, vol. 9408, pp. 18–31. Springer, Cham (2015). https://doi.org/10.1007/978-3-319-25645-0_2
29. Xiao, J., Xiao, Z., Wang, D., Bai, J., Havyarimana, V., Zeng, F.: Short-term traffic volume prediction by ensemble learning in concept drifting environments. Knowl. Based Syst. **164**, 213–225 (2018)

A Feature Selection Method for Multi-dimension Time-Series Data

Bahavathy Kathirgamanathan[✉] and Pádraig Cunningham

School of Computer Science, University College Dublin, Dublin, Ireland
bahavathy.kathirgamanathan@ucdconnect.ie

Abstract. Time-series data in application areas such as motion capture and activity recognition is often multi-dimension. In these application areas data typically comes from wearable sensors or is extracted from video. There is a lot of redundancy in these data streams and good classification accuracy will often be achievable with a small number of features (dimensions). In this paper we present a method for feature subset selection on multidimensional time-series data based on mutual information. This method calculates a merit score (MSTS) based on correlation patterns of the outputs of classifiers trained on single features and the 'best' subset is selected accordingly. MSTS was found to be significantly more efficient in terms of computational cost while also managing to maintain a good overall accuracy when compared to Wrapper-based feature selection, a feature selection strategy that is popular elsewhere in Machine Learning. We describe the motivations behind this feature selection strategy and evaluate its effectiveness on six time series datasets.

Keywords: Time-series classification · Feature selection · Merit score

1 Introduction

Multi-dimension time-series data arises in various application areas such as motion capture and activity recognition [9,11]. This data will often contain a lot of redundancy with some of the data streams being highly correlated. For this reason, it is important to be able to identify a subset of the features (data streams) that is adequate to characterize the phenomenon under investigation. This is a special case of the feature selection problem in Machine Learning (ML) but in this case the 'feature' is a complete time-series rather than a feature in a feature vector representation.

Time-series data is often not compatible with the standard ML feature selection strategies. Filter strategies are not directly applicable due to the nature of the data and Wrapper methods can be computationally prohibitive (see Sect. 2 for more detail).

In this paper, a feature subset selection method for multivariate time series is implemented with the aim of identifying the optimal feature subset to use for classification. The method uses feature-feature correlations as well as feature-class correlations based on mutual information (MI) which are then used to

© Springer Nature Switzerland AG 2020
V. Lemaire et al. (Eds.): AALTD 2020, LNAI 12588, pp. 220–231, 2020.
https://doi.org/10.1007/978-3-030-65742-0_15

calculate a merit score for each feature subset which will act as the basis upon which to select the 'best' subset. The main novelty is that these correlations are calculated on the outputs of classifiers trained on single features rather than on the time-series data.

The following section of this paper presents an overview of existing feature selection techniques. Section 3 describes the Merit Score based technique used for time series (MSTS), Sect. 4 presents our evaluation of MSTS on selected datasets, and finally Sect. 5 discusses the conclusions and scope for further work.

2 Feature Selection

In a data set of n dimensions there are 2^n possible feature subsets. Feature Selection techniques explore this space of feature subsets to find the 'best' subset. Evaluation strategies can be divided into two broad categories:

- **Filter** methods use an external measure such as information gain or a χ^2 statistic to score the informativeness of features. Then a selection criterion will determine the best features to select according to this score, e.g. select features scoring above a threshold or select the top m features.
- **Wrapper** methods for feature selection make use of the learning algorithm itself to choose a set of relevant features. The Wrapper conducts a search through the feature space, evaluating candidate feature subsets by estimating the predictive accuracy of the classifier built on that subset. The goal of the search is to find the subset that maximises this criterion.

Filter methods are not computationally expensive but are less accurate as features are not evaluated in context. Wrapper methods can be very effective because they evaluate what is important, the classification performance of different feature subsets. However, because of the extent and nature of the evaluation, Wrappers are computationally expensive.

2.1 Correlation Based Feature Selection Using Mutual Information

Correlation based feature selection (CFS) is a compromise between Filter and Wrapper methods as it evaluates features in context but using correlation rather than classification accuracy [5]. CFS is the default feature selection method in Weka [4] and has been widely used. However, CFS is not usable with time-series data because it requires data in a feature vector format. CFS assigns a merit score M_S to a feature subset as follows:

$$M_S = \frac{k\overline{r_{cf}}}{\sqrt{k + k(k-1)\overline{r_{ff}}}} \tag{1}$$

Where $\overline{r_{cf}}$ is the average correlation between the features in the subset and the class label and $\overline{r_{ff}}$ is the average correlation between the selected features. k represents the number of features in the subset. These correlations have been measured using techniques such as symmetrical uncertainty based on information gain, feature weighting based on the Gini-index, and a method using the minimum description length (MDL) principle [5]. Information gain based methods have worked well previously and hence the correlations in this paper will be measured using Mutual Information (MI). MI has been widely used and has produced successful results for feature selection [3]. Generally, as the MI between two random variables increases, the greater the correlation between them will be.

MI is a concept that is used widely in information theory and is based on Shannon's entropy [13], which is a measure of the uncertainty of random variables. Given two continuous random variables X and Y, the entropy of X is defined as:

$$H(X) = -\int p(x)\, log\, p(x)\, dx \tag{2}$$

The entropy of X and Y is defined as:

$$H(X,Y) = -\iint p(x,y)\, log\, p(x,y)\, dx\, dy \tag{3}$$

The MI between X and Y is defined as:

$$MI(X;Y) = \iint p(x,y)\, log\, \frac{p(x,y)}{p(x)p(y)}\, dx\, dy \tag{4}$$

where $p(x,y)$ is the joint probability density function of X and Y and $p(x)$ and $p(y)$ are the probability density function of X and Y respectively.

Hence MI and entropy can be combined in the form:

$$MI(X;Y) = H(X) + H(Y) - H(X,Y) \tag{5}$$

In this paper, the adjusted mutual information (AMI) score is used to calculate the correlations. The AMI score is an adjustment of the MI score to account for chance [14]. The AMI score is defined as:

$$AMI(X,Y) = \frac{MI(X,Y) - E[MI(X,Y)]}{mean(H(X),H(Y)) - E[MI(X,Y)]} \tag{6}$$

2.2   Feature Selection for Time-Series Data

A time series is a time based sequence of observations, $x_i(t); [i = 1, \ldots, n; t = 1, \ldots, m]$, where i indexes the data gathered at time point t. The time series is univariate when n is 1 and multivariate when n is greater than or equal to 2. Multivariate time series can often be large in size and hence it is important to have suitable methods for preprocessing the data prior to classification.

To deal with the high dimensionality of MTS, two common methods used are feature extraction and feature subset selection. Feature extraction methods involve the transformation or mapping of the original data into extracted features. Feature subset selection involves reducing the number of features from the original dataset that is used for analysis by selecting only the features required and removing the redundant features. One potential downfall of using feature extraction methods is that there can be a loss of information compared to using the original features. In this paper, the focus will be on feature subset selection methods.

Many state of the art feature subset selection techniques such as Recursive Feature Elimination (RFE) require each item to be inputted in the form of a column vector [8]. Multivariate time series tend to naturally be represented as a $m \times n$ matrix which makes these methods not ideal when working with multivariate time series for correlation based feature selection as vectorising time series data will lead to a loss of information about the correlation between the features. Hence, although there has been a lot of work undertaken in the area of multiple variable feature selection, there is limited work in feature selection for multivariate time series (MTS).

Some correlation based methods have been implemented for feature subset selection in time series. Many methods typically used to calculate correlation such as Spearsman's correlation and rank correlation can be effective for non-time series data however has been shown to produce poor results when implemented on time series [15].

Principal Component Analysis (PCA) is another technique that has been used in multivariate feature selection which allows correlation information between variables to be preserved. CLeVer is a technique which utilises properties of the descriptive common principal components for MTS feature subset selection. This method uses loadings to weight the contribution of each feature to the principal components. By ranking each feature by how much it contributes to the principal components, this method aims to reduce the dimensionality while retaining information related to both the original features and the correlation amongst the features [16].

Mutual Information (MI) is a popular technique that has been used on MTS data to measure correlation. MI is advantageous over other methods as it allows for both linear and nonlinear correlation to be captured. The class separability based feature selection (CSFS) algorithm uses MI between the original variables as features for classification. Based on this, the ratio of between class scattering to within class scattering is used to identify the contribution of a feature to the classification, hence allowing the original variables to be ranked according to their contribution to the classification [6].

MI is generally calculated in a pairwise manner which may not be ideal when working with multidimensional data. To avoid this, some studies have used a k-nearest neighbour (k-NN) approach to calculate the MI which avoids the need to calculate the probability distribution function and therefore can be used on the original multidimensional feature subset [7,10]. Many of the methods using MI select the feature one by one using greedy search methods which may not lead to the identification of the optimal subset. The MSTS approach taken in this paper uses MI to evaluate correlation which is then used to calculate a merit score for each subset from which the best subset is selected.

3   CFS for Time-Series Data

CFS relies on the principle that "*a good feature subset is one that contains features highly correlated with the class, yet uncorrelated with each other*" [5]. In this context where we use time series data, we aim to find a subset with features which are good predictors of the class while sharing little information with the other features in the subset.

Typically, the correlation between the feature values themselves are calculated for use in CFS. As this is not feasible with time series data, we use the predictions of the class labels from each feature to help identify which features may be more correlated. The correlations could be defined in various ways including any distance measure between the feature-class and feature-feature class label predictions or through the use of mutual information based approaches. While investigating the best method to use to measure the correlations, initially the single feature accuracy was used for feature-class correlation and Hamming distance was used for feature-feature correlations. However, we decided to take a mutual information based approach for the correlations instead as it proved to give better accuracy.

(I) Make predictions using single feature

	0	1	2	3	4	...	298	299
TRUE Predictions (F_{true})	2	2	5	3	6	...	4	3
Feature 1 Predictions (F_1)	2	3	5	3	6	...	4	4
Feature 2 Predictions (F_2)	2	2	1	4	6	...	4	4
Feature 3 Predictions (F_3)	2	2	5	1	6	...	5	3
Feature 4 Predictions (F_4)	2	2	5	5	6	...	4	4

(II) Calculate feature-feature correlations and feature-class correlation using AMI score (Equation 6)

AMI(F_x,F_y)	F_1	F_2	F_3	F_4	F_{true}
F_1		0.19	0.17	0.19	0.65
F_2			0.18	0.20	0.57
F_3				0.23	0.61
F_4					0.61

(III) Calculate Merit score using Equation 7 where k = 2

Merit Score(F_x,F_y)	F_1	F_2	F_3	F_4
F_1		0.79	0.82	0.82
F_2			0.77	0.76
F_3				0.78
F_4				

Fig. 1. Process taken to calculate the merit score of the ERing dataset for the 2-feature subset case. This is a six class problem with class labels 1 to 6.

Figure 1 shows the process followed to calculate the MSTS where we initially make a prediction using each of the features separately. The predictions are then compared with the true labels using the Adjusted Mutual Information (AMI) score (Eq. 6) to find the feature-class correlations and compared with the predictions of the other features, again using AMI to find the feature-feature correlations (Fig. 1 (II)). Once the correlations have been identified, the merit score can be calculated using a modified version of Eq. 1 as follows:

$$MSTS = \frac{k\overline{Y_{cf}}}{\sqrt{k + k(k-1)\overline{Y_{ff}}}} \tag{7}$$

where Y_{cf} and Y_{ff} are correlations calculated on the class labels predicted for the training data rather than on feature values as is the case in Eq. 1. Hence, Y_{cf} was calculated by averaging the feature-class AMI score of all the features present in the subset. Y_{ff} is calculated as the average of the pairwise AMI scores of each combination of features in the subset. The 'best' subset would ideally be the one with the largest merit score. In this example there is a tie between F_1, F_3 and F_1, F_4. - see Fig. 1 (III).

Following the merit score calculation, further evaluation is required to select the 'best feature subset'. To do this, we evaluate two strategies. The strategies taken were:

1. **Strategy 1.** The merit scores are calculated for all possible feature subset combinations (see Sect. 4.2). The feature subset with the highest merit score was selected as the best feature subset.
2. **Strategy 2.** Merit scores are calculated as in Strategy 1. The top 5% of the merit scores were selected and a Wrapper search was carried out on the selected feature subsets to identify the feature subset with the highest accuracy.

Further detail on how this algorithm was evaluated is presented in the following section.

4   Evaluation

In our evaluation we aim to assess the effectiveness of MSTS to identify good performing feature subsets and investigate how efficient this approach would be in terms of computational cost.

4.1   Data Sets

Six datasets were used for evaluation and these were all taken from the UEA multivariate time series classification archive [1]. Five of these datasets are related to activity recognition and motion capture with one dataset from the audio spectra domain. All datasets were selected to have four or more dimensions. Four of the six datasets consist of accelerometer and/or gyroscope data. The ArticularyWordRecognition dataset has data obtained from an electromagnetic articulograph, a small sensor placed on the tongue and the JapaneseVowels dataset was taken from audio recordings. A summary of the datasets used for evaluation is shown in Table 1.

Table 1. Summary table of datasets used for evaluation

	Total # of samples	# of classes	# of variables	Time series length
ArticularyWordRecognition (AWR)	575	25	9	24
JapaneseVowels (JW)	640	9	12	29
Cricket (Cr)	180	12	6	1197
ERing (ER)	60	6	4	65
NATOPS (NT)	360	6	24	51
RacketSports (RS)	303	4	6	30

4.2   Merit Score Evaluation

The evaluation of the merit score was undertaken for feature subsets up to and including 4 features which was deemed sufficient as often MTS data only requires a small number of features to obtain high accuracy. To calculate the merit score for each dataset the following steps were taken:

1. **Identify all unique feature subsets**. All unique combinations of feature subsets was identified and stored.
2. **Calculate and store DTW distance matrix**. The similarity measure used for the time series in this paper is Dynamic Time Warping (DTW). DTW allows for a mapping of the time series in a non-linear way and works to find the optimal alignment between both series. DTW can be considered as a one-to-many mapping [12]. As this is a computationally expensive task and will be repeatedly used for cross-validation, it is calculated and stored in advance.
3. **Make class label predictions for each feature**. A 1-NN classifier using the stored DTW distances was used to do a 3-fold cross validation to make a set of class predictions using each feature individually.
4. **Calculate feature-class and feature-feature correlations**. Calculate the feature-feature correlations and feature-class correlations as explained in Sect. 3.
5. **Calculate Merit Scores using** Eq. 7.

To compare the effectiveness of the merit score in identifying the optimal subsets, the classification accuracy of each subset was also calculated using a 1-NN-DTW classifier, which is often used as a benchmark technique whilst working with time series [2]. A 3-fold cross validation was performed for each dataset. Figure 2 shows the merit score against its subset accuracy for each feature subset.

A positive trend is seen in Fig. 2 where a higher merit score generally corresponds to a higher accuracy. This trend is very visible in five out of six of the datasets with the NATOPS dataset yielding a less promising correlation in comparison with the other datasets. This behaviour may be due to the innate characteristics of the data which suggests that this approach may be more suitable for some datasets and domains than others. A slight feature subset size bias (SS-bias) is seen in the datasets where the different subsets sizes are forming clusters. However, overall the merit score gives a good indication of the better performing feature subsets and if the highest merit score was selected, a subset with a good classification accuracy would be selected as the 'best' subset, although the optimal subset may not be selected. This is further evaluated in Sect. 4.3

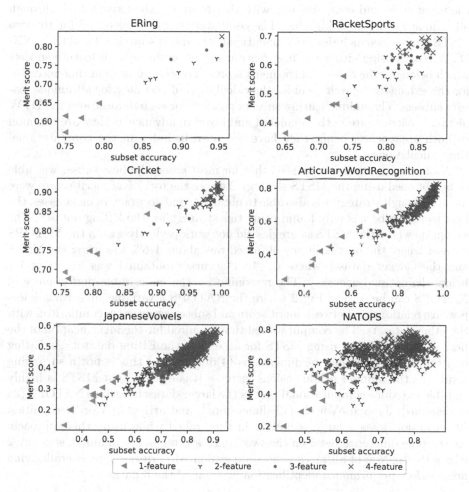

Fig. 2. MSTS vs subset accuracy for each feature subset combination for the six datasets

4.3   Feature Subset Selection

Following the merit score calculation, the two strategies where we take the highest merit score to represent the best feature subset (Strategy 1) and we take the top 5% of merit scores and undertake a search through this to find the best feature subset (Strategy 2) were both implemented on the datasets. For all evaluations of performance a 3-fold cross validation using 1-NN-DTW was used on the selected feature subsets. Figure 3 shows a comparison between the best accuracy and computational time required by the two strategies undertaken using the merit score and compares this with that from an exhaustive search through all unique feature combinations. The computational time recorded for the two MSTS approaches includes the calculation of the merit score itself and the 1-NN-DTW search using either the best feature subset or through all feature subsets which belong to the top 5% of the merit scores. The computational time recorded for the exhaustive search includes the calculation of accuracy for all unique feature subsets. The unique feature subsets possible for each dataset and the DTW distance matrices are both calculated and saved in advance as they are common to both approaches, hence they have not been included in the computational time calculations.

From the results it can be seen that for most cases, the best subset was able to be obtained using the MSTS strategy 2 where the top 5% of merit scores were used. Although strategy 1 is also able to obtain a good accuracy in most cases, the best feature subset is only found using this strategy for the ERing dataset. The exception where the MSTS strategies did not work perfectly was in the NATOPS dataset where the best accuracy obtained was about 4–5% less using the MSTS and the Cricket dataset where the best accuracy obtained was less than 1% below the optimal accuracy. The reasoning for the undesirable performance of NATOPS can be seen in Fig. 2 where the NATOPS 4 variable subset has a less positive relationship between merit score and subset accuracy in comparison with the other datasets. The computational time required for the identification of the best subset was faster using MSTS for all except the ERing dataset. As ERing had the smallest number of dimensions (4 dimensions) this is not a surprising result. As the number of dimensions increase it is evident that MSTS is highly suitable to reduce computational cost as the larger datasets such as NATOPS (24 dimensions), JapaneseVowels (12 dimensions), and articularyWordRecognition (9 dimensions) see a large reduction in time taken while using this approach. As the time difference between the two MSTS strategies are minimal, strategy 2 where the top 5% of all merit scores are evaluated performs best overall giving near perfect performance identification in 5 out of the 6 datasets.

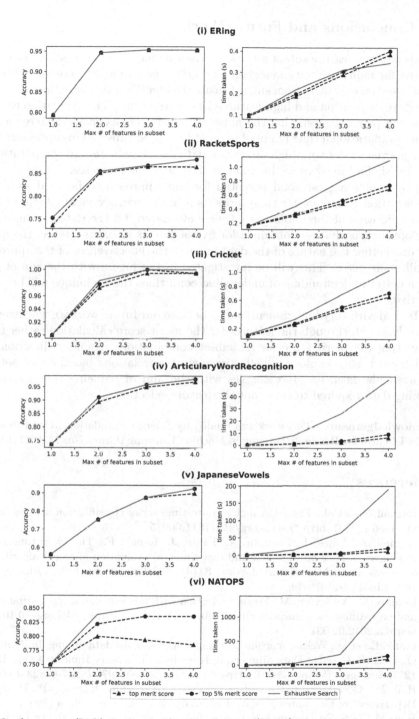

Fig. 3. Accuracy (Left) and computational time (Right) required for each of the datasets using the two MSTS strategies vs an exhaustive search

5 Conclusions and Future Work

In this paper, a feature subset selection technique based on merit scores is implemented for multivariate time series. The technique employed here uses correlations based on classifiers from single features to identify a subset with low feature to feature correlation and high feature-class correlations. The evaluation carried out in this paper suggests that this approach can lead to a considerable reduction in the computational time required to identify a good subset. This approach is in particular useful for very high dimension data as the reduction in computational time by MSTS improves as the number of dimensions increases.

The results suggest good potential for this approach to be used as a feature selection technique for time series as a high accuracy yielding subset was selected in each of the datasets that were evaluated. Of the datasets analysed, near optimal results were obtained for five of the six datasets. Hence, the question of whether the nature of the data impacts the effectiveness of the approach is still unanswered. This will be investigated in the future with the aim of getting a better understanding of under what conditions this technique will be most effective.

To deal with very high dimension datasets, in our future work we will attempt a greedy search through the features for the merit score calculation rather than calculating the merit score for all subset combinations. Another direction for further work is to explore the effectiveness of correlations based on subsets of the available data, e.g. 100 samples with the aim of reducing the amount of training data required to carry out the feature selection.

Acknowledgements. This work was funded by Science Foundation Ireland through the SFI Centre for Research Training in Machine Learning (Grant No. 18/CRT/6183).

References

1. Bagnall, A., et al.: The UEA multivariate time series classification archive, 2018, pp. 1–36 (2018). http://arxiv.org/abs/1811.00075
2. Bagnall, A., Lines, J., Bostrom, A., Large, J., Keogh, E.: The great time series classification bake off: a review and experimental evaluation of recent algorithmic advances. Data Min. Knowl. Discov. **31**(3), 606–660 (2016). https://doi.org/10.1007/s10618-016-0483-9
3. Doquire, G., Verleysen, M.: Feature selection with missing data using mutual information estimators. Neurocomputing **90**, 3–11 (2012). https://doi.org/10.1016/j.neucom.2012.02.031
4. Frank, E., et al.: Weka-a machine learning workbench for data mining. In: Maimon, O., Rokach, L. (eds.) Data Mining and Knowledge Discovery Handbook, pp. 1269–1277. Springer, Boston (2009). https://doi.org/10.1007/978-0-387-09823-4_66
5. Hall, M.: Correlation-based feature selection for machine learning. Ph.D. thesis, Department of Computer Science, University of Waikato Hamilton (1999)
6. Han, M., Liu, X.: Feature selection techniques with class separability for multivariate time series. Neurocomputing **110**, 29–34 (2013). https://doi.org/10.1016/j.neucom.2012.12.006

7. Han, M., Ren, W., Liu, X.: Joint mutual information-based input variable selection for multivariate time series modeling. Eng. Appl. Artif. Intell. **37**, 250–257 (2015). https://doi.org/10.1016/j.engappai.2014.08.011

8. Isabelle Guyon, A.E.: An introduction to variable and feature selection. J. Mach. Learn. Res. **3**, 1157–1182 (2003). https://doi.org/10.1016/j.aca.2011.07.027

9. Johnston, W., O'Reilly, M., Coughlan, G., Caulfield, B.: Inertial sensor technology can capture changes in dynamic balance control during the Y balance test. Digit. Biomark. **1**(2), 106–117 (2018). https://doi.org/10.1159/000485470

10. Liu, T., Wei, H., Zhang, K., Guo, W.: Mutual information based feature selection for multivariate time series forecasting. In: Chinese Control Conference, CCC 2016, 7110–7114, August 2016. https://doi.org/10.1109/ChiCC.2016.7554480

11. O'Reilly, M., Caulfield, B., Ward, T., Johnston, W., Doherty, C.: Wearable inertial sensor systems for lower limb exercise detection and evaluation: a systematic review. Sport. Med. **48**(5), 1221–1246 (2018)

12. Sakoe, H., Chiba, S.: Dynamic programming algorithm optimization for spoken word recognition. IEEE Trans. Acoust. Speech Signal Process. **26**(1), 43–49 (1978)

13. Shannon, C.E.: A mathematical theory of communication. Bell Syst. Tech. J. **27**(4), 623–656 (1948). https://doi.org/10.1002/j.1538-7305.1948.tb00917.x

14. Vinh, N.X., Epps, J., Bailey, J.: Information theoretic measures for clusterings comparison: variants, properties, normalization and correction for chance. J. Mach. Learn. Res. **11**, 2837–2854 (2010)

15. Wang, Q.G., Li, X., Qin, Q.: Feature selection for time series modeling. J. Intell. Learn. Syst. Appl. **05**(03), 152–164 (2013). https://doi.org/10.4236/jilsa.2013.53017

16. Yang, K., Yoon, II., Shahabi, C.: CLeVer: a feature subset selection technique for multivariate time series (full version). Technical report (2005)

Author Index

Printed in the United States
By Bookmasters